The CREATION of BRIDGES

The CREATION of BRIDGES

From vision to reality - the ultimate challenge of architecture, design, and distance

David Bennett

CHARTWELL
BOOKS, INC.

A QUINTET BOOK

Published by Chartwell Books
A Division of Book Sales, Inc.
114, Northfield Avenue
Edison, New Jersey 08837

This edition produced for sale in the U.S.A., its territories and
dependancies only.

ISBN 0-7858-1053-6

This book was designed and produced by
Quintet Publishing Limited
6 Blundell Street
London N7 9BH

Creative Director: Richard Dewing
Designer: James Lawrence
Senior Editor: Sally Green

Typeset in Great Britain by
Central Southern Typesetters, Eastbourne
Manufactured in Hong Kong by
Regent Publishing Services Ltd
Printed in China by
Leefung-Asco Printers Ltd

Picture Credits

Key Numbers refer to page numbers. *b* = bottom, *l* = left, *m* = middle, *r* = right, *t* = top.

Margaret Amman-Durer 201*t*, 202*b*; **Boilly Photo, Prince Edward Island, Canada** 222*tl*; **British Cement Association** 55*mr*, **Caltrans** 47*t*, 136, 152; **Julie V Clark** 41*tr*, 63, 64*bl*, 66, 67, 81*r*, 87, 154, 216*b*; **Construction News** 173; **Robert Cortright** 15*ml*, 14*m*, 17, 20*b*, 21, 25*m*, 34*tr*, 41*ml*, 43*br*, 53, 55*t*, 57*bl*, 71*t*, 101, 104, 105, 134, 157, 169, 184, 187, 189*t*, 192*t*; **Jenny Crossley** 22*b*, 23, 109*m*, 143, 225*t*; **Eric Delony** 27*b*, 29, 35, 41*b*, 43*br*, 44*t*, 64*bl*, 71*b*, 77*b*, 137, 167, 201*b*, 205; **ETH Zurich** 73; **FBM Studio** 55; **Hugh Ferguson** 141*t,* 171; **Flint & Neil Partnership** 52*tl*, 65*r*, 140, 153, 170, 172; **Dave Freider** 39, 128, 202*t*, 207; **Freyssinet** 54, 57*br*, 70*b*, 74*m*, 75*t* & *r*, 79*m*, 121, 122, 208, 209, 210, 211, 212; **Freyssinet/photographer Francis Vigouroux** 28, 52*b*, 58*b*, 59, 64*b*, 79*b*, 80, 93, 94, 95, 96, 97, 127, 145*t*, 146, 148, 149, 150, 151; **Jolyon Gill** 31*t*, 186; **Honshu-Shikoku Bridge Authority** 49*tr*, 81*l*; **Institution of Civil Engineers** 9, 11, 18, 27*t*, 32, 33, 99, 108, 109*t*, 110, 111, 112, 113, 118, 119, 120, 138, 139*t*, 158, 159, 161, 162, 163, 165, 166, 168, 177, 178, 179, 180, 181*b*, 182, 183, 185, 188, 189*b*, 190*t*, 191*t*, 192*t*, 193; **Jean Muller International** 57, 58*tl* & *r*, 98, 142, 144, 175, 219, 220, 221, 222, 223, 224, 225*b*, 226*t*, 227; **Fritz Leonhardt** 30*m* & *b*, 49*tl*, 51, 52*mr*, 56*tr*, 57*ml*, 64*tl*, 65*tl*, 190*b*, 191*b*, 213*b*, 214*b*, 215, 216*t*, 217*t*, 218; **Life File** 197; **MTA Bridges & Tunnels Special Archive** 50, 90*bl*, 91*r*; **J-L Michotey** 16, 22*t*, 44*b*, 72*t*, 77*tr*, 103, 109*b*, 115, 181; **New York Port Authority** 47*br*, 129, 203, 204; **Ove Arup & Partners** 56*b*; **Jorg Schlaich** 10; **Grant Smith** 19, 41*tl*, 123, 130, 131, 132, 133, 226*b*; **Alain Spielmann** 74*b*, 106, 107, 116; **Steinman Consulting Engineers** 36, 37, 38, 41, 42*tl* & *bl*, 43*tr*, 45, 46*bl*, 47*bl*, 48, 49*m*, *b*, 76, 78*tl*, *tr*, 83, 85, 88*b* & *r*, 89*r*, 90*tl*, *br*, 135; **Denton Taylor** 174, 193, 195, 194, 197, 200; **Michel Virlogeux** 13, 79*tr*, 114; **Hans Wittfoht** 8, 12*t*, 14*t*, 42*bl*, 64*mr*, 100*b*, 214*t*.

Contents

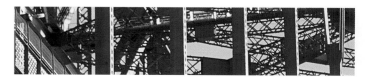

Foreword

Bridges have been and still are the umbilical cords of humankind's progress over the centuries. From the simple log and the crude rope suspension bridge to the monumental concrete and steel spans of today, bridges have dramatically unfurled the history of humanity's conquest of nature's barriers—a river, a chasm, an estuary, a valley, and even the sea. Bridges just a few feet wide to many miles long have the same common goal of serving the need for better access, better transportation links and trade between local communities and international boundaries.

In the past century there has been more bridge building, more discovery of new technology and advances in bridge-engineering science than in all the preceding centuries put together. The year 1998 has been a bumper year for bridge-building activity, on a scale so vast and so financially huge that it is difficult to comprehend. In Japan alone the current bridge-building program is the equivalent of building all of New York's bridges in one go. Just imagine the Brooklyn, the George Washington, the Verazzano Narrows, the Bronx Whitestone, Throgs Neck, Manhattan, Williamsburg, Queensboro, the Goethals, Hellgate, Bayonne, and Outerbridge all built one after the other and within ten years. It beggars belief, but it is happening.

Why this urgency? The answer is space for housing for a rapidly increasing population, and for the future needs for commerce and transportation, between the Honshu mainland and the island of Shikoku. The Japanese government are building an infrastructure that must serve the whole community and stimulate economic growth in Japan for the next hundred years.

In charting the history of bridge building through the centuries, recalling some of the greatest bridges ever built and the horrific tales of the worst bridge tragedies, one thing is dominant and common among them all. No matter how big or how small they may be, every bridge project must start with a vision of its creation, followed by the endeavor to make that creation a reality. Bridges were not built as monuments for pleasure or grandeur but as an economic necessity in the service of a community or nation. They blazed a trail over inhospitable lands, over rapid-flowing streams and deep gorges.

This book explains by simple diagrams and everyday experiences the fundamental principles behind the design and construction of a bridge, from post-and-lintel and arch-and-truss bridges to box girders, cable-stay, and suspension bridges. The many illustrations and examples of the different types of bridges will help you to recognize them during your travels. The stunning locations and elegance of many of the bridges shown may also inspire you to make a special journey or detour while on vacation or business to see some of these man-made wonders.

You will see many unfamiliar terms as you read on. To help, I have provided a glossary at the back, which explains most of the terms. Others should become apparent from context.

I have one or two regrets in researching my material: I could not find space to put everything in. I have not been able to include material about footbridges, swing bridges, lifting bridges, and transporter bridges, nor say much about the current controversy in the bridge world over who should conceptually design a bridge—an architect or an engineer.

I wish to acknowledge the help I received from Freyssinet International with copies of their publications and the many images of bridges; Jean Muller International (JMI) for information and the many images on JMI bridges; Michael Chrimes at the ICE Library in London; Eric DeLony of HAER; George Gesner of Steinman Consulting Engineers for archive material; Alain Spielmann, Michel Virlogeux, Jean Muller, Fritz Leonhardt, and Angus Low for their views on bridges.

The inspiration for this book has been sourced from David Steinman's wonderful book *Bridges and Their Builders*, Joseph Gies's *Bridges and Men*, Hans Wittfoht's highly informative *Bridge Building* and Mario Salvadori's thought-provoking *Building—from caves to skyscrapers*.

David Bennett, October 1998

The early history of bridge building
–the age of timber and stone

The bridge has been a feature of human progress and evolution ever since man the hunter-gatherer became curious about the world beyond the horizon and the fertile land, the animals and fruit flourishing on trees on the other side of a river or gorge. Early human beings had to devise ways to cross a stream and a deep gorge to survive.

A boulder or two dropped into a shallow stream works well as a stepping stone, as many of us have discovered—but, for deeper-flowing streams, a tree dropped between banks is a more successful solution. So the primitive idea of a simple beam bridge was born.

Although it is certain that early humans lived in groups and passed on such primitive technology, it is likely that the skill of making rudimentary bridges was discovered and rediscovered by succeeding generations many times over, until it became established as a skill. For those groups who lived in the forests, surviving on fruit, grubbing up roots, and snaring or spearing animals, ease of travel through the forest canopy was vital to locate new food sources and shelters.

ABOVE: A rope suspension bridge in Asia.

OPPOSITE: Primitive log bridge, Afghanistan.

Today, in the forests of Peru and the foothills of the Himalayas, crude rope bridges span deep gorges and fast-flowing streams to maintain pathways from village to village for hill tribes. Such primitive rope bridges evolved from the vine and creeper that early humans would have used to swing through the forest and to cross a stream. We have only to see a gibbon on a wildlife program or a chimpanzee at play in a zoo to get the idea of how it would work. Here is the second basic idea of a bridge: the suspension bridge.

For thousands of years during the Paleolithic period (2,000,000–10,000 BC), we know that our ancestors lived as nomads and wanderers, hunting and gathering food. Slowly it dawned on early man that following herds of deer or buffalo—or just foraging for plant food haphazardly—could be better managed if the animals were kept in herds nearby and plants were grown and harvested in fields. Regular routes between settlements became necessary to barter for grain and food stocks—and invariably to use when stealing from each other! If there was a shorter way to travel between two places, human enterprise would find a way to bridge a river or to cut a clearing through a forest.

In this period the simple log bridge had to serve many purposes. It needed to be broad and strong enough to take cattle; it needed to be a level and solid platform to transport food and other materials; and it needed to be movable so that it could be withdrawn to prevent enemies from using it.

Narrow tree trunk bridges were inadequate and were replaced by double-log beams spaced wider apart on which short lengths of logs were placed and tied down to create a pathway. The pathways were planed by sharp scraping tools or axes in the Bronze Age and any gaps between them plugged with branches and earth to create a level platform.

For crossings over wide rivers, support piers were formed from piles of rocks in the stream. Sometimes stakes were driven into the river bed to form a circle and then filled with stones, creating a crude cofferdam—a watertight, dry enclosure. Around 3500 BC early Bronze Age "lake dwellers" lived in timber houses built out over the lakes, in the area that is now Switzerland. To ensure the house did not sink, they evolved ways to drive timber piles into the lake bed. From the discovery of this came the timber-pile bridge and the trestle bridge.

For tribal groups living in the more northern glacial regions, with a plentiful supply of stones of all shapes and lengths and not as many large trees, the stone slab was preferred as a beam to bridge rivers and streams. Crude stone bridges that survive today in Dartmoor, England, and built around the fifteenth century, are reminiscent of prehistoric stone-bridge construction of the Bronze Age.

So primitive bridges were essentially post-and-lintel structures, made either from timber or stone or from a combination of both. Sometime later the simple rope-and-bamboo suspension bridge was devised, which developed into the rope suspension bridges that are in regular use today in the mountain reaches of China, Peru, Colombia, India, and Nepal.

It took human ingenuity till about 4000 BC to discover the secrets of arch construction. In the Tigris–Euphrates valley the Sumerians began building with adobe—a sun-dried mud brick—for their palaces, temples, ziggurats, and city defenses. Stone was not plentiful in this region and had to be imported from Persia, so was used sparingly.

BELOW: An engraving of an early Indian pontoon bridge.

The brick module dictated the construction principles they employed, to scale any height and to bridge any span. And, through trial and error, it was the arch and the barrel vault that were devised to build their monuments and grand architecture at the peak of their civilization. The ruins of the magnificent barrel-vaulted brick roof at Ptsephon and the Ishtar Gate at Babylon are a reminder of Mesopotamian skill and craftsmanship.

Although most Egyptian building was dominated by post-and-lintel stone construction, the corbeled arch had been discovered and was used frequently in constructing passageways, relieving arches, and escape tunnels within massive pyramid structures. At Dindereh today three arch ruins still stand that date from 3600 BC. By the end of the Third Dynasty, around 2475 BC, the Egyptians had mastered the true arch and it was in common use in construction.

Without doubt the arch is one of the greatest discoveries. The arch principle was the vital element in all building and bridge technology over later centuries. Its dynamic and expressive form gave rise to some of the greatest bridge structures ever built.

Earliest records of bridges

The earliest written record of a bridge appears to be one built across the Euphrates around 600 BC as described by Herodotus, the fifth-century Greek historian. The bridge linked the palaces of ancient Babylon on either side of the river. It had 100 stone piers, which supported wooden beams of cedar, cypress, and palm to form a carriageway 35 feet wide and 600 feet long. Herodotus mentions that the floor of the bridge would be removed every night as a precaution against invaders.

In China, it seems that bridge building evolved at a faster pace than the ancient civilization of Sumeria and Egypt. Records exist from the time of Emperor Yoa in 2300 BC on the traditions of bridge building.

ABOVE: Cantilever timber bridge in Tibet.

BELOW: Illustration of a covered cantilever truss bridge in medieval China, which became the focal point for trade in a town.

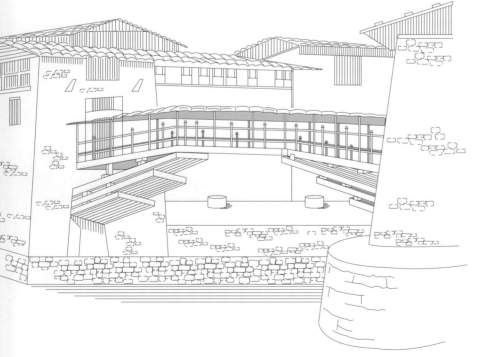

Early Chinese bridges included pontoons or floating bridges and probably looked like the primitive pontoon bridges built in China today. Boats called sampans, about 30 feet long, were anchored side by side in the direction of the current and then bridged by a walkway.

The other bridge forms were the simple post-and-lintel beam, the cantilever beam, and rope suspension cradles. Timber beam bridges, probably like those of Europe, were often supported on rows of timber piles of soft fir wood called "foochow poles," so named because they were grown in Foochow. A team of builders would hammer the poles into the river bed using a cylindrical stone fitted with bamboo handles. A short crosspiece was fixed between pairs of poles to form the supports that carry timber boards, which were then covered in clay to form the pathway over the river.

For longer spans and for bridges that must maintain a wide navigation channel, the cantilever beam or rope suspension bridge was the usual choice. The cantilever bridge, being rigid in construction, was preferred in the towns and cities and for river crossing along important trading routes.

To build a cantilever bridge, first a wooden caisson—a structure for keeping water out while deep foundations are being excavated—was formed on each bank and filled with stone, rubble and clay. Then timbers were driven into the front of the caisson and embedded deep in the fill to form a platform of cantilevers that spring from each bank. The two arms of the cantilever were then bridged over the central gap by simple beams.

As stone cutting and masonry construction was better understood, the wooden caisson was replaced by a stone abutment, which was often built into a house or gateway structure leading onto the bridge span. If a central pier was needed, it was often adorned with a pagoda or canopy structure to serve as a meeting place and an open market for buying and selling goods. The town bridge over a wide river became the town square in Chinese society and the center for commerce and trade.

In later centuries Chinese bridge building was dominated by the arch, which they copied and adapted from the Middle East as they traveled the silk routes that opened during the Han Dynasty around AD 100.

Through Herodotus we learn about the Persian ruler Xerxes (c. 519–465 BC) and the vast pontoon bridge he had built, consisting of two parallel rows of 360 boats, tied to one another and to the bank and anchored to the bed of the Hellespont, which is the Dardanelles today. Xerxes wanted to get his army of 2 million fighting men and horses to the other side to meet the Greeks at Thermopylae. It took seven days and seven nights to get the army over to the other side.

Xerxes' massive army was defeated at the Battle of Thermopylae in 480 BC, the remnants of which retreated back over the pontoon bridge to fight another day. The Persians were great bridge builders and built many arch, cantilever, and beam bridges. There is a bridge still standing at Dizful in Khuzistan, Iran, over the river Diz, which could date anywhere from 350 BC to AD 400. The bridge consists of 20 voussoir arches (formed from wedge-shaped stones) which are slightly pointed, hence the Gothic affinity, and has a total length of 1,250 feet. Above the level of the arch springing are small semicircular spandrel arches, which give the entire bridge an Islamic look, hence the uncertainty of its Persian origins.

The Greeks did not do much bridge building in their illustrious history, being a seafaring nation that lived on self-contained islands and in feudal groups scattered across the Mediterranean. They used exclusively post-and-lintel construction in

BELOW: Kintai bridge, Japan, built in 1673, is typical of many early Chinese and Japanese arch bridges.

evolving a classical order in their architecture, and built some of the most breathtaking temples, monuments, and cities the world has ever seen, such as the Parthenon, the Temple of Zeus, the city of Ephesus, Miletus, and Delphi, to name but a few. They were quite capable of building arches like their forebears the Etruscans when they needed to. There are examples of Greek voussoir arch construction that compare to the Beehive Tomb at Mycenea, like the ruins of an arch bridge with a 27-foot span at Pergammon in Turkey.

The Romans

The Romans on the other hand were the masters of practical building skills. They were a nation of builders, who took arch construction to a science and high art form during their domination of Mediterranean Europe. Their influence on bridge-building technology and architecture has been profound.

TOP: Partially destroyed stone arch bridge in Iran (Persia), built by the Sassanids (c400 AD).

ABOVE: Roman Aqueduct of Segovia in Spain, built during the end of the first century AD.

The Romans conquered the world as it was known then, built roadways, canals, and cities that linked Europe to Asia and North Africa, and produced the first true bridge engineers in the history of civilization. Romans understood that the establishment and maintenance of their empire depended on efficient and permanent communications. Building roads and bridges was therefore a high priority.

The Romans also realized, as did the Chinese in later centuries, that timber structures, particularly those embedded in water, had a short life and were prone to decay, insect infestation, and fire hazards. Prestigious buildings and important bridge structures were therefore built in stone. But the Romans had also learned to preserve their timber structures by soaking timber in oil and resin as a protection against dry rot, and coating them with alum for fireproofing. They learned that hardwood was more durable than softwood, and that oak was best for substructure work in the ground, alder for piles in water; while fir, cypress, and cedar were best for the superstructure above ground.

They understood the different quality of stone that they quarried. Tufa, a yellow volcanic stone, was good in compression but had to be protected from weathering by a stucco—a lime wash. Travertine was harder and more durable and could be left exposed, but was not very fire-resistant. The most durable materials, such as marble, had to be imported from distant regions of Greece and even as far as Egypt and Asia Minor (Turkey). The Romans' big breakthrough in material science was the discovery of lime mortar and pozzolanic cement, which was based on the volcanic clay that was found in the village of Pozzuoli. They

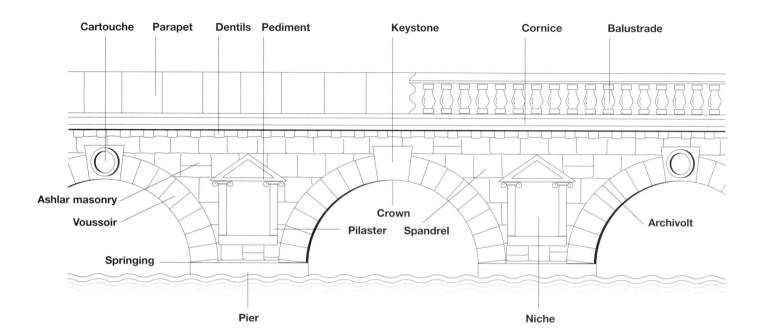

Cartouche Parapet Dentils Pediment Keystone Cornice Balustrade

Ashlar masonry
Voussoir

Crown
Pilaster Spandrel

Archivolt

Springing

Pier Niche

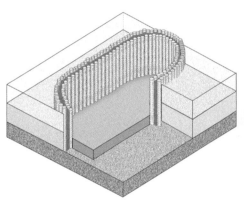

ABOVE: Architectural terms used in Roman and later masonry arch construction.

ABOVE: Roman coffer dam formed by two rows of timber piles, which are in-filled with clay to create a watertight enclosure.

ABOVE LEFT: Pons Fabricus in Rome, built in 62 BC. The modern name of the bridge is Ponte Quattro Capi, named after the branch of the Tiber that it spans.

used it as mortar for laying bricks or stones and often mixed it with burnt lime and stones to create a waterproof concrete.

The Romans realized that voussoir arches could span farther than any unsupported stone beam, and would be more durable and robust than any other structure. They ought to have known because the early Roman leaders and settlers were Etruscans. Semicircular arches were always built by the Romans, with the thrust from the arch going directly down onto the support pier. It meant that piers had to be large. If they were built wide enough at about-one third of the arch span, then any two piers could support an arch without shoring or propping from the sides.

In this way it was possible to build a bridge from shore to shore, a span at a time, without having to form the entire substructure across the river before starting the arches. They developed a method of constructing the foundation on the river bed within a cofferdam or watertight, dry enclosure, formed by a double ring of timber piles and clay packed into the gap between them to act as the water seal. The water inside the cofferdam was then pumped out and the foundation substructure was built within it. The massive piers often restricted the width of the river channel, increasing the speed of flow past the piers and increasing the scour action. To counter this the piers were built with cutwaters, which were pointed to cleave the water so it would not scour the foundations.

The stone arch was built on a wooden framework built out from the piers and known as centering The top surface was shaped to the exact semicircular profile of the arch. Parallel arches of stones were placed side by side to create the full width of the roadway. The semicircular arch meant that all stones were cut identically and that no mortar was needed to bind them together once the keystone was locked in position. The compression forces in the arch ensured complete stability of the span.

Of course the Romans did build many timber bridges but they have not stood the test of time, and today all that remains of their achievement after 2,000 years is a handful of stone bridges in Rome, and a few scattered examples in France, Spain, North Africa, Turkey, and other former Roman colonies. But what still stand today, whether bridges or aqueducts, rank among the most inspiring and noble bridge structures ever built, considering the limitation of their technology.

BELOW: Pont du Gard, Nîmes, France (c20 AD).

In Rome there are six bridges to be seen, the most celebrated of which is the highly decorated Ponte Sant'Angelo (136 BC) although many would regard the Pons Augustus in Rimini (AD 20) as the finer example, because of its classic proportions. Of the great aqueducts there is the Segovia Aqueduct (AD 1) in Spain, with its two tiers of 109 arches carrying the Rio Frio the last 2,500 feet into the town, and the most famous one of them all, the Pont du Gard at Nîmes, built in 19 BC. Its scale and monumental presence continue to impress.

The Dark Ages and the Brothers of the Bridge

When the Roman Empire collapsed it seemed that the light of progress around the world went out for a long while. The Huns, the Visigoths, Saxons, Mongols, and Danes did not do much building in their raids across Europe and Asia, plunder and destruction being higher on their agenda. It was left to the spread of Christianity and the strength of the church to start the next boom in road building and bridge building around AD 1000. It was the church that had preserved and developed both spiritual understanding and the practical knowledge of building during this period. This building naturally included the building of bridges.

ABOVE: The remains of Pont d'Avignon over the Rhône at Avignon, in France (1188).

In northern Italy there lived a group of friars of the Altopascio order near Lucca, in a large dwelling called the Hospice of St James. The friars were skilled at carpentry and masonry, having built their own priory. The surrounding countryside was wild and dangerous, and the refuge they built was a popular resting place for pilgrims and travelers using the ancient road from Tuscany to Rome.

In 1244 Emperor Frederick II required that the hospice build a proper bridge across the White Arno for pilgrims and travelers. Obviously the Hospice of St James was profiting well from passing trade for the Emperor to issue such a decree. With their skills they set up a cooperative to build the bridge.

After they had completed the bridge over the White Arno their fame spread through Italy and France. It sparked off an interest in bridge building among other ecclesiastical orders. In France a group of Benedictine monks established the religious order of the Frères Pontiffes (Brothers of the Bridge) to build a bridge over the Durance.

ABOVE: Engraving of "Old London Bridge" in the seventeenth century.

And so the Brothers of the Bridge order became established among Benedictine monks and spread from France to England by the thirteenth century. The purpose of the order, apart from its spiritual duties, was to give aid to travelers and pilgrims, to build bridges along pilgrim routes or to establish boats for their use, and to receive them in hospices built for them on the bank. The Brothers of the Bridge were great teachers, who strove to emulate and continue the magnificent work of the Roman bridge builders.

The most famous and legendary bridge of this period was built by the Order of Saint Jacques du Haut Pas, whose great hospice once stood on the banks of the Seine in Paris on the site of the present church with that name. They built the Pont Esprit over the Rhône, but their masterpiece was the neighboring bridge at Avignon. It was truly a magnificent and record-breaking achievement for its time. Its beauty has inspired writers, poets, and musicians over the centuries. What, then, was so special about the bridge? For a start the arch was not semicircular but elliptical in shape and therefore could span farther than a semicircular arch. It was more stable and could be made more slender over the crown. The result of all this was that the piers could be made narrower and the arch taller, thereby carrying the roadway higher out of reach of potential flooding and better for navigation. Small relieving arches were formed above the piers and in the spandrels to accommodate spring flood waters.

To have designed such a bridge you would have to be a mathematical wizard or have received divine inspiration from somewhere. The Pont d'Avignon comes with a legend about its shepherd-boy designer, Bénézet, who is said to have had a vision from God in 1178 commanding him to build it. When the Bishop of Avignon demanded proof of the boy's claim to divine intervention, the boy allegedly miraculously picked up an enormous boulder and carried it to the place where the bridge was to be built.

ABOVE: Ponte Vecchio in Florence, Italy (c1345).

The sanguine explanation is that Pont d'Avignon was masterminded by Brother Benoit, who supervised the brothers in the building of the Saint Esprit bridge and many other bridges. For its overall length of 1,300 feet and 20 spans, the bridge was remarkably slender and took only ten years to build, which is fast by medieval standards. The width of the bridge was also interesting. At its widest point it was 16 feet, but where the chapel was built over the second pier it narrows down to just 6 feet 6 inches. Such bottlenecks were designed by medieval bridge builders in order to defend the bridge more easily or perhaps to ensure a pilgrim toll was paid.

Sadly all that remains today at Avignon are just four out of the 20 spans of the bridge and the chapel where the supposed creator of the bridge was interred and later canonized as Saint Bénézet.

While Pont d'Avignon was being built in France, another monk of the Benedictine order in England, Peter of Colechurch, was planning the building of the first masonry bridge over the Thames. A campaign for funds was launched with enthusiasm. Not only did the rich townspeople, the merchants, and money lenders make generous donations, but the common people of London all gave freely. Until the sixteenth century a list of donors could be seen hanging in the chapel on the bridge. The structure that was built in 1206 was Old London Bridge and ranks after Pont d'Avignon in fame. It was such a popular bridge that

Development of arch construction

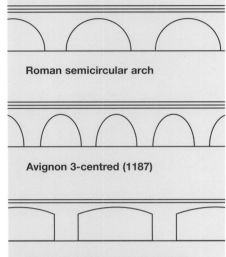

Roman semicircular arch

Avignon 3-centred (1187)

Ponte Vecchio segmental (1345)

buildings and warehouses were soon erected on it. It became so fashionable a location that the young noblemen of Queen Elizabeth's household resided in a curious four-story timber building imported piece by piece from Holland, called the "Nonesuch House."

Towns continued to sponsor and promote the building of stronger and better bridges and roads. They did not always get the Brothers of the Bridge to build them, because they were often committed to other projects for many years. Instead, guilds of master masons and carpenters were formed and spread across Europe offering their services. Even government officials were united in this community enterprise and began to grasp the initiative and drive for better road and bridge networks across the country.

Soon the vestiges of the Dark Ages and feudalism were transformed to the Age of Enlightenment and the Renaissance. The Ponte Vecchio in Florence built toward the end of this period marks the turning point of the Dark Ages. It was a covered bridge erected in 1345, lined with jewelers' shops and galleries, with an upper passageway added later, that was a link between the royal palaces and those of government—the Uffizi and Pitti. The piers, which are 20 feet thick, support the overhanging building as well as the bridge spans. The most innovative features of the bridge are the arch spans, which are extremely shallow compared with any previous arches ever built or indeed many contemporary European bridges. It was built as a segmental arch, which is unusual for bridge builders of that period because they could not possibly determine the thrust from the arches mathematically, with the knowledge they had. How they did it is a mystery, as is the segmental arch of Pont d'Avignon. The architect of this radical design was Taddeo Gaddi, who had studied under the great painter Giotto, and was regarded as one of the great names of the Italian Renaissance that followed.

RIGHT: Example of a medieval fortified bridge. Monmow Bridge, Monmouth, Wales (1272).

The Renaissance

Not since the days of Homer, Aristotle, and Archimedes in Hellenistic times have such great feats of discovery in science and mathematics, and such works of art and architecture, been achieved as during the Renaissance. Modern science was born in this period through the inquiring genius of Copernicus, Leonardo da Vinci, Francis Bacon, and Galileo and in art and architecture through Michelangelo, Brunelleschi, and Palladio. During the Renaissance there was a continual search for the truth, explanations of natural phenomena, greater self-awareness, and rigorous analysis of Greek and Roman culture. As far as bridge building was concerned, particularly in Italy, it was regarded as a high art form.

As much emphasis was placed upon its decorative order and pleasing proportions as on the stability and permanence of its construction. Bridge design was architect-driven for the first time with da Vinci, Palladio, Brunelleschi, and even Michelangelo all experimenting with the possibility of new bridge forms. The most significant contribution of the Renaissance was the invention of the truss system, developed by Palladio from the simple king-post and queen-post roof truss, and the founding of the science of structural analysis with the first book ever written on the subject by Galileo Galilei entitled

Dialoghi delle Nuove Scienze (*Dialogues on the New Science*) published in 1638.

Palladio did not build many bridges in his lifetime; many of his truss bridge ideas were considered too daring and radical and his work lay forgotten until the eighteenth century. His great treatise published in 1520—*Four Books Of Architecture*, in which he applied four different truss systems for building bridges—was destined to influence bridge builders in future years when the truss replaced the Roman arch as a principal form of construction.

Other groups of bridge builders during the Renaissance were clever material technologists who were preoccupied with the art of bridge construction and how they could build with less labor and materials. It was a time of inflation when the price of building materials and labor was escalating. The most famous bridge builders in this era were Ammannati, Da Ponte, and Du Cerceau.

Which bridge of the Renaissance is the most beautiful? Santa Trinità in Florence? The Rialto in Venice? The Pont Neuf in Paris? Arguably the most

famous and celebrated bridge of the Renaissance was the Rialto bridge designed by Antonio Da Ponte in Venice . "The best building raised in the time of the Grotesque Renaissance, very noble in its simplicity, in its proportions and its masonry." So said John Ruskin about the Rialto. Its designer was 75 years old when he won the contract to build the Rialto, and was 79 when it was finished. It is a single segmental arch span of 87 feet 7 inches, which rises 25 feet 11 inches at the crown. The bridge is 75 feet 3 inches wide, with a central roadway, shops on both sides, and two small paths on the outside, next to the parapets. Two sets of arches, six each of the large central arch, support the roof and enclose the 24 shops within it. It took three and a half years to build, and all the city's stonemasons were kept busy for two years.

It was also of a very novel construction, as you can see in Chapter 4, where I've gone into some detail in the section that looks at my pick of what I consider to be some of the best bridges in the world.

Equally innovative and skillful bridge construction was being progressed across Europe. In the state of Bohemia across the Moldau at Prague was built the longest bridge over water, the Karlsbrücke, in 1503, which was the most monumental and imperial bridge of the Renaissance. It took a century and half to complete. It was adorned with statues of saints and martyrs and terminates on each bank with an imposing tower gateway.

In France during this time, a fine example of the early French Renaissance, the Pont Neuf, was being designed. It was the second stone bridge to be built in Paris and, although its design and construction did not represent a

TOP: Pont Neuf, Paris, France (c1607).

ABOVE: The semi-circular arches of the Pont Neuf.

great leap forward in bridge building, it occupies a special place in Parisian hearts. It was designed by Jacques Androuet du Cerceau, and its two arms, which join the Ile de la Cité to the left and right bank of the Seine, represented a massive undertaking. Although all the arches are semicircular and not segmental, no two spans are alike, as they vary from 31 to 61 feet in span and also differ on the downstream and upstream sides of each arch, which were built on a skew of 10 percent. Du Cerceau wanted the bridge to be a true unencumbered thoroughfare bereft of any houses and shops. But the people of Paris demanded shops and houses and got their way in the end, which resulted in modification to the few short-span piers that had been constructed.

The Pont Neuf has stood now for 400 years and was the center of trade, and the principal access to the crowded island when it was built. The booths and stalls on the bridge became so popular that all sorts of traders used it, including booksellers, pastry cooks, jugglers, and peddlers. They crowded the roadway until there were some 200 stalls and booths packed into every niche along the pavement.

The longer left bank of the Pont Neuf was extensively reconstructed in 1850 to exactly the same details, after many years of repairs and attention to its poor foundations. The right bank with the shorter spans has been left intact. The entire bridge has been cleared of all stalls and booths and is used today as a road bridge.

The finest examples of late French Renaissance bridge building during the seventeenth century were the Pont Royal and Pont Marie bridges, which are still standing today. The Pont Royal was the first bridge in Paris to feature elliptical arches and the first to use an open caisson to provide a dry working place in the river bed. The foundations for the bridge piers were designed and constructed under the supervision of Francosi Romain, a preaching brother from Holland who was an expert in solving difficult foundation problems. The bridge architect, François Mansart, and the builder, Jacques Gabriel, called on Romain after they ran into foundation problems.

Romain introduced dredging in the preparation of the river bed for the caisson using a machine that he had developed. After excavations were finished the caisson was sunk to the bed, but the top was kept above the water level. The water was then pumped out and the masonry work of the pier was built inside the dry chamber. The five arch spans of the Pont Royal increase in span toward the center and, although the bridge has practically no ornamentation, it blends beautifully into its river setting.

The Renaissance brought improvement in both the art and science of bridge building. For the first time people began to regard bridges as civic works of art. The master bridge builder had to be an architect, structural theorist, and practical builder, all rolled into one.

The bridge that was without doubt the finest exhibition of engineering skills in this era was the slender, elliptical-arched bridge of Santa Trinità at Florence, designed by Bartolommeo Ammannati, in 1567. Many scholars are still mystified to this day as to how Ammannati arrived at such pleasing slender, curves to the arches.

TOP: Elliptical arches of Pont Royale in Paris (1687).

ABOVE: Pont Marie, Paris (seventeenth century).

The eighteenth century: The Age of Reason

In this period masonry arch construction reached perfection, owing to a momentous discovery by Jean Perronet and the innovative construction techniques of John Rennie. Just as the masonry arch reached its zenith 7,000 years after the first crude corbeled arch in Mesopotamia, it was to be threatened by a new building material—iron—and the timber truss, as the principal construction for bridges in the future.

This was the era when the civil engineering as a profession was born, when the first school of engineering was established in Paris at the Ecole de Paris during the reign of Louis XV. The director of the school was Jacques Gabriel, who had designed the Pont Royal. He was given the responsibility of collecting and assimilating all the information and knowledge there was on the science and history of bridges, buildings, roads, and canals.

ABOVE: Rennie's "New London Bridge" under construction.

With such a vast bank of collective knowledge, it was inevitable that building architecture and civil engineering should be separated into the two fields of expertise. It was suggested it was not possible for one man in his brief life to master the essentials of both subjects. Moreover, it became clear that the broad education received in civil engineering at the Corps des Ponts et Chaussées (or Bridge and Highway Corps) at the Ecole de Paris was not sufficient for the engineering of the large projects. More specialized training was needed in bridge engineering. In 1747 the first school of bridge engineering was founded in Paris at the historic Ecole des Ponts et Chaussées. The founder of the school was Trudiane, and the first teacher and Director was a brilliant young engineer named Jean-Rodolphe Perronet.

Jean Perronet has been called the "father of modern bridge engineering" for his inventive genius and design of the greatest masonry arch bridges of the century. In his hand the masonry arch reached perfection. The arch he chose was the curve of a segment of a circle of larger radius, instead of the familiar three-centered arch. To express the slenderness of the arch he raised the haunch of the arch considerably above the piers.

Perronet was the first person to realize that the horizontal thrust of the arch was carried through the spans to the abutments, and that the piers, in addition to carrying the vertical load, also had to resist the difference between adjacent span thrusts. He deduced that, if the arch spans were about equal and all the arches were in place before the centering was removed, the piers could be greatly reduced in size.

What remains of Perronet's great work? Only his last bridge, the glorious Pont de la Concorde in Paris, built when he was in his eighties. It is one of the most slender and daring stone-arch constructions ever built. "Even with modern analysis," suggests Professor James Finch, the author of *Engineering and Western Civilization*, "we could not further refine Perronet's design."

With France under the inspired leadership of Gabriel and then Perronet, the rest of Europe could only admire and copy these great advances in bridge building. In England a young Scotsman named John Rennie was making his mark, following in the footsteps the great French engineers. He was regarded as the natural successor to Perronet, who was a very old man when Rennie began his career. Rennie was a brilliant mathematician, a mechanical genius and pioneering civil engineer. In his early years he worked for James Watt to build the first steam-powered grinding mills at Abbey Mills in London, and later designed canals and drainage systems to drain the marshy fens of Lincolnshire.

Rennie built his first bridge in 1779 across the Tweed at Kelso. It was a modest affair with a pier-width-to-span ratio of 1:6 and with a conservative elliptical arch span. He picked up the theory of bridge design from textbooks and from studies and discussion about arches and voussoirs with his mentor Dr Robison of Edinburgh University. He designed bridges with a flat, level roadway and not the characteristic hump of most English bridges. It was radical departure from convention and was much admired by all the townsfolk, farmers, and traders who transported material and cattle across them.

This bridge was a modest forerunner to the many famous bridges that Rennie went on to build—including Waterloo, Southwark, and New London Bridge. What was Rennie's contribution to the *science* of bridge building then? For Waterloo bridge, the centering for the arches was assembled on shore then floated out on barges into position. So well and efficiently did this system work that the framework for each span could be put in position in a week. This was a fast erection speed and, as a result, Rennie was able to halve bridge construction time.

So soundly were Rennie's bridges built that 40 years later Waterloo bridge had settled only 5 inches. Rennie's semi-elliptical arches, sound engineering methods and rapid assembly technique, together with Perronet's segmental arch, divided pier, and understanding of arch thrust, changed bridge design theory for all time.

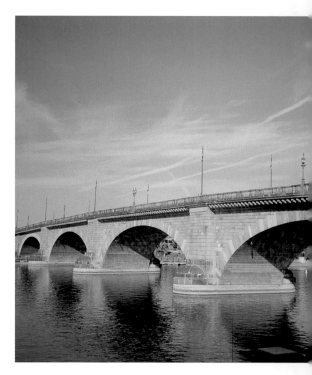

TOP: "New London Bridge" was completed in 1831.

ABOVE: Rennie's bridge was rebuilt at Lake Havanasu, Arizona in 1972.

The carpenter bridges

America with her vast expansion of roads and waterways, following her commercial growth in the eighteenth century, was to become the home of the timber bridge in the nineteenth.

America had no tradition or history of building with stone, and so early bridge builders used the most plentiful and economical material available: wood. They produced some of the most remarkable timber bridge structures ever seen, but they were not the first to pioneer such structures. The Grubenmann brothers of Switzerland were the first to design a quasi-timber truss bridge in the eighteenth century. The Wettingen bridge over the Limmat just west of Zürich was considered their finest work. The bridge combines the arch-and-truss principle with seven oak beams bound closely together to form a catenary arch to which a timber truss was fixed. The span of the Wettingen was 309 feet and far exceeded the span of any other timber bridge at the time. Palmer, Wernwag, and Burr—the so called carpenter-bridge builders of North America, who designed more by intuition than by calculation—developed the truss arch to span farther than any other wooden construction. This was the last of the three basic bridge forms to be discovered. The first man who made the truss arch bridge a success in America and who patented his truss design was Timothy Palmer, a New England Yankee. In 1792 he built a bridge consisting of two trussed arches over the Merrimac; similar to one of Palladio's truss designs, except the arch was the dominant supporting structure.

Palmer's "Permanent Bridge" over the Schuylkill, built in 1806, was his most celebrated bridge. When the bridge was finished the president of the bridge company suggested that it would be a good idea to cover it to preserve the timber from rot and decay in the future. Palmer went further than that and timbered the sides as well, completely enclosing the bridge. Thus America's distinctive covered bridge was established. By enclosing the bridge it stopped snow getting in and piling up on the deck, thus causing it to collapse from the extra load.

Wernwag was a German immigrant from Pennsylvania, who built 29 truss-type bridges in his lifetime. His designs integrated the arch and truss into one composite structure rather more successfully than Palmer's.

Wernwag's famous bridge was the "Colossus" over the Schuylkill just upstream from Palmer's "Permanent Bridge" and comprised two pairs of parallel arches, linked by a framing truss, which carried the roadway. The truss itself was acting as bracing reinforcement and consisted of heavy verticals and light diagonals. The diagonal elements were remarkable because they were iron rods, and were the first iron rods to be used in a long-span bridge. In its day the Colossus was the longest wooden bridge in America, having a clear span of 304

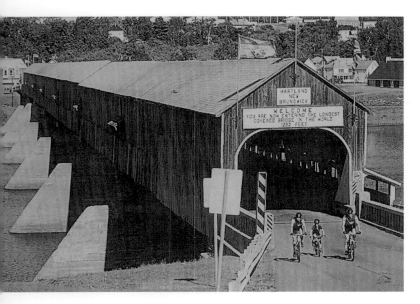

TOP: The covered bridge at Hartland, New Brunswick. As it proudly states, it is the world's longest covered bridge.

ABOVE: Illustration of the Wettingen bridge, Switzerland.

feet. Fire destroyed the bridge in 1838. It was replaced with Charles Ellet's pioneering suspension bridge.

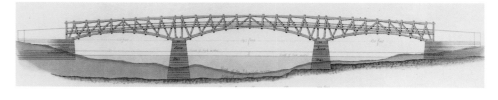

Theodore Burr was the most famous of the illustrious triumvirate. Burr developed a timber truss design based on the simple king-and-queen-post truss of Palladio. He came closest to building the first true truss bridge, but it proved unstable under moving loads. Burr then strengthened the truss with an arch. It was significant that here the arch was added to the truss rather than the other way round.

ABOVE: Palmer's "Permanent Bridge" over the Schuykll River, USA.

Burr's arch trusses were quick to assemble and modest in cost to build, and for a time they were the most popular timber bridge form in America. By 1820 the truss principle had been well explored and, although the design theory was not understood, in practice it had been tested to the limit.

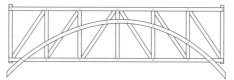

The Burr arch/truss (1815)

It was left to Ithiel Town to develop and build the first true truss bridge, which he patented and called the Town Lattice. It was a true truss because it was free from arch action and any horizontal thrust. Simple to build, it could be nailed together in a few days and was cheap compared with other options. Town promoted his bridge with the slogan "built by the mile and cut by the yard."

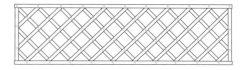

The Town lattice truss (1820)

With the arrival of the railroad in the US, bridge building continued to develop along two separate ways. One school continued to evolve stronger and leaner timber truss structures, while the other experimented with cast and wrought iron.

BELOW: "Bollman Truss Bridge" (1869), a famous iron bridge in Savage, Maryland, USA.

The first patent truss to incorporate iron into a timber structure was the Howe Truss—named after William Howe, who pioneered the development of truss bridges in the US. It had top and bottom chords and diagonal bracing in timber and the vertical members made of iron rods in tension. This basic design was, with modification, to last right into the next century.

The first fully designed truss was the Pratt Truss, which reversed the forces of the Howe Truss by putting the vertical timber members in compression and the iron diagonal members in tension. In 1847, the Whipple truss (named after the American civil engineer Squire Whipple) was the first all-iron truss—a bowstring truss—with the top chord and vertical compression members made from cast iron and the bottom chord and diagonal bracing members made from wrought iron.

Later Fink, Bollman, Bow, and Haupt in the US, along with Cullman and Warren in Europe, developed the truss to a fine art, incorporating wire-strand cable, timber, and iron to form lightweight, strong bridges to carry railroads.

A pictorial history of modern bridge building—the last 200 years

Bridge development in the nineteenth and twentieth centuries

The Industrial Revolution, which began in Britain at the end of the eighteenth century, gradually spread throughout the world and brought with it huge changes in all aspects of everyday living. New forms of bulk transportation, by canal and rail, were developed to keep pace with the increasing exploitation of coal and the manufacture of textiles and pottery. Coal fueled the hot furnaces to provide the high temperatures to smelt iron. Henry Bessemer invented a method to produce crude steel alloy by blowing hot air over smelted iron. Siemens and Martins refined this process further to produce the low-carbon steels of today.

High temperature was also essential in the production of cement, which John Aspen discovered by burning limestone and clay on his kitchen stove in Leeds, England, in 1824. Wood and stone were gradually replaced by cast-iron and wrought-iron construction, which in turn was replaced by first steel and then concrete, the two primary materials of bridge building in the twentieth century.

ABOVE: The steel-arched St Louis bridge, USA (1874).

OPPOSITE: The cable-stayed Coatzacoalas 11 Bridge, Mexico (1984).

"Old London Bridge" fifteenth century; stone pointed arch.

"New London Bridge" 1831; stone segmental arch.

"London Bridge" 1968; prestressed concrete flat arch.

ABOVE: Progress of arch construction from the fifteenth to the twentieth century.

BELOW: Brooklyn Bridge, New York, 1993.

BOTTOM: Brooklyn Bridge, New York, 1858.

Growing towns and expanding cities demanded continuous improvement and extension of the road, canal, and railroad infrastructure. The machine age introduced the steam engine, the internal-combustion engine, factory production lines, domestic appliances, electricity, gas, processed food, and the tractor.

Faster assembly of bridge building was essential, and this meant prefabricating lightweight but tough bridge components. The heavy steam engines and longer goods train imposed larger stresses on bridge structures than ever before. Bridges had to be stronger and more rigid in construction and yet had be faster to assemble to keep pace with progress. Connections had to be stronger and more efficient. The common nut and bolt were replaced by the rivet, which was replaced by the high-strength friction-grip bolt and the welded connection.

When the automobile arrived, it resulted in a road network that eventually crisscrossed the entire countryside from town to city, over mountain ranges, valleys, streams, rivers, estuaries, and seas. Even bigger and better bridges were now needed to connect islands to the mainland and countries to continents, to open up major trading routes. The continuous search for and development of high-strength materials such as steel, concrete, carbon fiber, and aramids have today combined with sophisticated computer analysis and dynamic testing of bridge structures against earthquakes, hurricane wind, and tidal flows to enable bridges to span even farther. In the last two centuries the bridge span has leapt from 350 to over 6,000 feet. This is the age of the mighty suspension bridges, the elegant cable-stayed

bridges, the steel-arch truss, the glued segmental and cantilever box-girder bridges.

The key events and achievements in this frenzied activity of bridge building are highlighted in pictorial form to illustrate the rapid pace of change and the many bridge ideas that were advanced. In the last two centuries more bridges were built than in the entire history of bridge building prior to that!

The age of iron (1775–1880)

Of all the materials used in bridge construction—stone, wood, brick, steel, and concrete—iron was used for the shortest time. Cast iron was first smelted from iron ore successfully by Dud Dudley in 1619. It was another century before Abraham Darby devised a method to economically smelt iron in large quantities. However, the brittle quality of cast iron made it safe to use only in compression in the form of an arch.

Wrought iron, which replaced cast iron many years later, was a ductile material that could carry tension. It was produced in large quantities after 1783 when Henry Cort developed a puddling furnace process to drive impurities out of pig iron.

But iron bridges suffered some of the worst failures and disasters in the history of bridge building. The vibration and dynamic loading from a heavy steam locomotive and goods wagons create cyclic stress patterns on the bridge structure as the wheels roll over the bridge, going from zero load to full load, then back to zero. Over a period of time these stress patterns can lead to brittle failure and fatigue in cast iron and wrought iron.

In one year alone in the US, as many as one in every four iron and timber bridges had suffered a serious flaw or had collapsed. Rigorous design codes, independent checking, and new bridge-building procedures were drawn up, but it was not soon enough to avert the worst disaster in iron-bridge history over the Tay Estuary in Scotland in 1879. It marked the end of the iron bridge for good.

ABOVE: Iron Bridge, Coalbrookdale, England (1779).

LEFT: Buildwas Bridge, Coalbrookdale, England (1796).

Landmarks of the age of iron

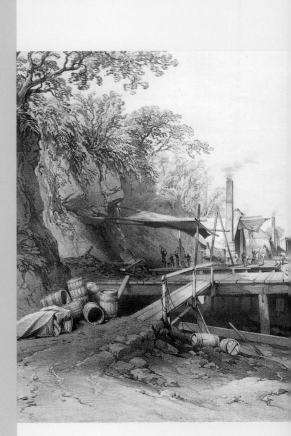

1779 **Coalbrookdale, England, the first cast-iron bridge, designed as an arch structure by Thomas Pritchard for its owner and builder, Abraham Darby III.**

1796 **Buildwas Bridge, the second cast-iron bridge built in Coalbrookdale, England, designed by Thomas Telford, used only half the weight of cast iron of Coalbrookdale.**

1807 **James Finlay builds the first elemental suspension bridge in wrought iron, the Chain Bridge, over the Potomac in Washington, DC.**

1821 **Guinless Bridge, England, George Stephenson's wrought-iron "lentilcular" girder bridge for the Stockton to Darlington Railway.**

1826 **Menai Straits Bridge, Wales, famous eyebar, wrought-iron chain-suspension bridge over the Menai Straits, by Thomas Telford.**

1834 **The Fribourg Bridge, Switzerland, the world's longest iron suspension bridge.**

1841 **Whipple patents the cast-iron "bowstring" truss bridge.**

1846 **Wheeling Suspension Bridge, USA, Charles Ellet's record-breaking, 1,000-foot-span, iron-wire suspension bridge.**

1850 **Britannia Bridge (Wales), first box-girder bridge concept, built in wrought iron by Robert Stephenson, son of George Stephenson.**

1853 **Murphy designs a wrought-iron Whipple truss, with pin connections.**

1858 **Royal Albert Bridge, Saltash, in the southwest of England: Isambard Kingdom Brunel's famous tubular-iron bridge over the Tamar.**

1876 **The Ashtabula Bridge disaster in USA: 65 people die when this iron, modified Howe truss collapses plunging train and passengers into the deep river gorge below.**

1879 **The Tay Bridge disaster, Dundee, Scotland, where a passenger train with 75 people on board plunges into the Tay estuary, as the supporting wrought iron girders collapse in high winds.**

FINLAYS "CHAIN SUSPENSION BRIDGE"

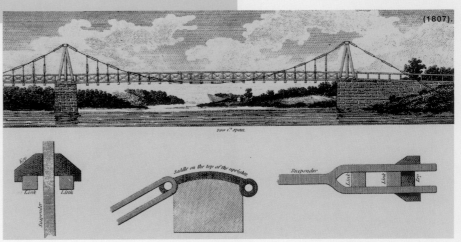

GARABIT VIADUCT, ST FLOUR, FRANCE (1884).

HELLGATE BRIDGE, NEW YORK, USA (1916).

SYDNEY HARBOUR BRIDGE, AUSTRALIA (1932).

NEW RIVER GORGE BRIDGE, USA (1978).

The cantilever truss Arch bridges had been constructed for many centuries in stone, then iron, and steel when it became available. Steel made it possible to build long-span trusses farther than cast iron, without any increase in the dead weight. Consequently, it made cantilever long-span truss construction viable over wide estuaries. The first and most significant cantilever truss bridge to be built was the railroad bridge over the Firth of Forth near Edinburgh, Scotland, in 1890. The cantilever truss was rapidly adopted for the building of many US railroad bridges until the collapse of the Quebec Bridge in 1907.

ABOVE: Forth Rail Bridge, Scotland (1890).

Landmarks

▓ 1886 **The Fraser River Bridge, Canada—believed to be the first balanced cantilever truss bridge to be built. All the truss piers, links, and lower chord members were fabricated from Siemens-Martin steel. It was dismantled in 1910.**

▓ 1890 **The Forth Rail Bridge, Edinburgh, Scotland—the world's longest spanning bridge at 1709 ft, when it was finished.**

▓ 1891 **The Cincinnati Newport Bridge, Cincinnati, USA—with its long, through, cantilever spans and short-truss spans—was the prototype of many railroad bridges in the USA.**

▓ 1902 **The Viaur Viaduct, France— this railroad bridge between Toulouse and Lyons was an elegant variation of the balanced cantilever, with no suspended section between the two cantilever arms.**

▓ 1918 **The Quebec Bridge, Canada— completion of the second Quebec bridge, the world's longest cantilever span.**

▓ 1927 **Carquinez Bridge—the last of the long cantilever truss bridges to be built in the US, although a second, identical, bridge was built alongside it in 1958 to increase traffic flow.**

THOUSAND ISLANDS BRIDGE—THE CANTILEVER
TRUSS SPAN—CANADA, USA (1938)

CARQUINEZ BRIDGE, USA (1927).

BRIDGE OF THE GODS, USA (1926).

POUGHKEEPSIE RAIL BRIDGE, USA (1888).

QUEBEC BRIDGE, CANADA (1918).

VIAUR VIADUCT, FRANCE (C1900).

The suspension bridge The early pioneers of chain suspension bridges were James Finlay, Thomas Telford, Samuel Brown, and Marc Seguin, but they had only cast and wrought iron available in the building of their early suspension bridges. It was not until Charles Ellet's Wheeling Bridge had shown the potential of the wire suspension principle using wrought iron that the concept was universally adopted. Undoubtedly, the greatest exponent of early wire suspension construction and strand-spinning technology was John Roebling. His Brooklyn Bridge was the first to use steel for the wires of suspension cables.

Suspension bridges are capable of huge spans, bridging wide river estuaries and deep valleys, and have been vital in establishing road networks across a country. They have held the record for the longest span almost unchallenged from 1826 to the present day—a record interrupted only between 1890 and 1928, when the cantilever truss held the record.

RIGHT: David Steinmann on the Brooklyn Bridge, after it was strengthened in 1953.

Landmarks—suspension bridges

▓ 1883 **Brooklyn Bridge, USA**—following the completion of the Wheeling suspension bridge pioneered by Charles Ellet; John Roebling went on to design the Brooklyn Bridge, the first steel-wire suspension bridge in the world.

▓ 1931 **George Washington, USA**—the heaviest suspension bridge to use parallel wire cables rather than rope-strand cables, and had the longest span in the world for nearly a decade.

▓ 1940 **Tacoma Narrows, USA**—the second Tacoma Narrows Bridge over Puget Sound in Washington State. This bridge, rebuilt after the collapse of the first bridge, with a deep stiffening truss deck, set the trend for future suspension bridge design in America.

▓ 1957 **Mackinac, USA**—Fondly referred to as "Big Mac" is the longest overall suspension bridge in America.

▓ 1964 **Verazzano Bridge, USA**—the last big suspension bridge to be built in America, and also held the record for the longest span until 1981.

▓ 1967 **Severn Bridge, England**—the first bridge to have a slim, aerodynamic bridge deck, eliminating the need for deep stiffening trusses like those of the American suspension bridges. It set the trend for future suspension-bridge construction.

▓ 1981 **Humber Bridge, England**—the longest span in the world when it was completed, with supporting strands that were inclined in zigzag fashion rather than the parallel arrangement preferred by the Americans.

▓ 1998 **Storebelt and the East Bridge, Denmark**—Storebelt is now substantially complete, and is the longest bridge in Europe. For a short while the main span of the East Bridge held the record for the longest span in the world.

▓ 1998 **Akashi Kaikyo, Japan**—is one of a family of long-span bridges linking the islands of Honshu and Shikoku, now well under construction. Its main span of 6,528 feet makes it the longest span in the world.

BROOKLYN BRIDGE, USA (1883).

LION'S GATE BRIDGE, VANCOUVER, BRITISH COLUMBIA, CANADA

CABLE ANCHORAGES OF THE THOUSAND ISLANDS SUSPENSION BRIDGE, CANADA, USA (1938).

OAKLAND BAY BRIDGE, SAN FRANCISCO, USA (1932).

ST JOHN'S BRIDGE, OREGON, USA (1931).

GEORGE WASHINGTON BRIDGE, USA (1931).

FROZEN WATERS UNDER THE MACKINAC BRIDGE.

TAGUS BRIDGE, LISBON, PORTUGAL (1966).

MACKINAC BRIDGE, USA (1957).

HUMBER BRIDGE, ENGLAND (1981).

ELEVATION OF THE TAGUS BRIDGE.

AKASHI KAIKYO BRIDGE, JAPAN (1998).

EAST BRIDGE, GREAT STOREBELT, DENMARK (1998).

VERAZZANO NARROWS BRIDGE, USA (1964).

Steel-plate-girder and box-girder bridges Since the development of steel and the I-beam, many beam bridges were built using a group of beams in parallel, which were interconnected at the top to form a roadway. They were quick to assemble but they were practical over only relatively short spans for rail and road viaducts. The riveted girder I-beam was later superseded by the welded and friction-grip bolted beam.

Relatively long spans were not efficient, however, because the depth of the beam needed could became excessive. To counter this, web plate stiffeners were added at close intervals to prevent buckling of the beam. Another solution was to make the beam into a hollow box, which was very rigid. In this way the depth of the beam could be reduced and material could be saved.

The steel box-girder beams were quick to fabricate and easy to transport. Their relatively shallow depth meant that high approaches were not necessary. Most of this pioneering work was carried out during and after World War II, when there was a huge demand for fast and efficient bridge building for spans of up to 1,500 feet. A major rebuilding program in Germany witnessed the construction of many steel box-girder and concrete box-girder bridges in the 1950s and 1960s. For spans greater than 1,500 feet the suspension and cable-stay bridge are generally more economical to build.

In the 1970s the world's attention was focused on the collapse of four steel box-girder bridges under construction. The four bridges were in Vienna over the Danube, in Milford Haven in West Wales (when four people were killed), a bridge over the Rhine in Germany, and the West Gate bridge in Melbourne, Australia, over the Lower Yarra River. By far the worst collapse was on the West Gate bridge, a single-cable-stay structure with a continuous box-girder deck. A cantilever section 376 feet long and weighing 1,200 tons buckled and crashed off the pier support onto some site huts 120 feet below, where workmen and engineers were having their lunch. Thirty-five people were killed in the tragedy. After that, further construction of steel box-girder decks was halted until better design standards, new site checking procedures, and a fabrication specification was internationally agreed.

BELOW: Elbe Bridge, Germany—steel plate girder (1936); Bonn-Buel Bridge, Germany—steel plate girder (1948).

Landmarks

- 1936 **Elbe Bridge—one of the early plate-girder bridges on the German autobahn.**

- 1948 **Bonn Beuel—a later development of the plate girder into a flat arch, to reduce material weight.**

- 1952 **Cologne Deutz Bridge—first slender, steel box-girder bridge in the world.**

- 1970s **Failure of box girders at Milford Haven in West Wales and West Gate Bridge in Australia stopped further development of the steel box-girder bridge decks.**

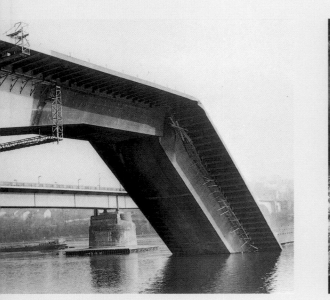

KOBLENZ BOX GIRDER
BRIDGE COLLAPSE, GERMANY.

BONN SOUTH RHINE BRIDGE IN GERMANY—STEEL BOX GIRDER (1967).

BOX GIRDER BRIDGES, A FEATURE OF MANY HIGHWAY INTERCHANGES.

Concrete and the arch (1900) Although engineers took longer to realize concrete's true potential as a material, it is used everywhere today in a vast array of bridges and building applications. Concrete is a brittle material just like stone, good in compression but not in tension, so if it starts to bend or twist it will crack. Concrete has to be reinforced with steel to give it ductility, so naturally its emergence followed the development of steel.

In 1824 Joseph Aspen made a crude cement from burning a mixture of clay and limestone at high temperature. The clinker that was formed was ground into a powder, and when this was mixed with water it reacted chemically to harden back into a rock. Nowadays, cement is usually combined with sand, stones, and water to create concrete, which remains fluid and plastic for a period of time, before it begins to set and eventually hardens. You can pour and place concrete into molds or forms while it is fluid, to create bridge beams, arch spans, support piers—in fact a variety of structural shapes. This gives concrete special qualities as a material, and scope for bold and imaginative bridge ideas.

ABOVE: Glenfinnan Viaduct, Scotland (1898).

François Hennebique was the first to understand the theory and practical use of steel reinforcement in concrete, but it was Robert Maillart (1872–1940) who was first to pioneer and build bridges with reinforced concrete. Eugene Freyssinet, Maillart's contemporary, was also keen to experiment with concrete structures. He went on to discover the art of prestressing and gave the bridge industry one of the most efficient methods of bridge deck construction in the world. Both these men were great engineers and champions of concrete bridges. What they achieved set the trend for future developments in concrete bridges—precast bridge beams, concrete arches, and box-girder and segmental cantilever construction. Concrete box-girder bridge decks are incorporated into many modern cable-stay and suspension bridges.

Hans Wittfoht, Jean Muller and the contractors Polensky and Zöllner, and Campenon Bernard, were responsible for building the first segmental and cast in place concrete box-girder bridges in the world. It is a technique that is used by many bridge builders across the world today. The box-girder span can be precast as segments or cast in place using a traveling formwork system. They can be built as a balanced cantilever each side of a pier or launched from one span to the next.

Concrete has been used in building most of the world's longest bridges. The relative cheapness of concrete compared with steel, the ability to rapidly precast or form prestressed beams of standard lengths, to bridge short spans over low-level trestle-type supports, had made concrete economically attractive. Lake Pontchartrain Bridge, a precast-concrete, segmental, box-girder bridge in Louisiana, is the longest bridge in the US with an overall length of 23 miles.

The Concrete Arch

▨ 1905 **Glenfinnan Viaduct, Scotland—** the first concrete arch bridge to be built in England.

▨ 1905 **Tavanasa Bridge, Switzerland—a breakthrough in the stiffened-arch slab.**

▨ 1922 **St Pierre de Vouvray, France—** early concrete bowstring arch of Freyssinet.

▨ 1930 **Salgina Gorge Bridge, Switzerland—one of the most aesthetic arch spans of Maillart.**

▨ 1930 **Plougastel Bridge, France—** unique construction concept which used prestressing for the first time.

▨ 1936 **Alsea Bay Bridge, USA—** completion of Conde McCullough's finest "art deco" bridge in concrete.

▨ 1964 **Gladesville, Australia—**use of precast, prestressed, segmental construction for the arch span. Also in 1964, KRK (Croatia)—the longest concrete arch span in the world.

GLADESVILLE BRIDGE, AUSTRALIA (1964).

PLOUGASTEL BRIDGE, FRANCE (1930).

KRK BRIDGE, CROATIA (1964).

SALGINA GORGE BRIDGE, SWITZERLAND (1930).

EARLY PICTURE OF THE SALGINA GORGE
BRIDGE, SWITZERLAND

SCHWANDBACH BRIDGE, SWITZERLAND (1933).

Concrete box girders In the 1950s and 1960s many freeway bridges and viaducts were built in Europe and USA using concrete box-girder construction. Some were precast segmental construction; some were cast in place.

FOLLOWING FOUR PAGE SPREAD:
General Belgrano Bridge over the Parana River in Argentina (1973)—cable stay navigation span, concrete box girder approach spans.

Landmarks

▨ 1952 **Shelton Road Bridge, USA—first match-cast, glued-segmental, box-girder construction in the world, developed by Jean Muller.**

▨ 1956 **Lake Pontchartrain Bridge, USA—the second longest bridge in the world. It is a precast, segmental, box-girder bridge with 2,700 spans and runs for 23 miles across Lake Pontchartrain near New Orleans. The second, identical, bridge, was built alongside the original one in 1969 and was just a little longer.**

▨ 1964 **Krahnenberg Bridge**

▨ 1972 **Medway Bridge, England—the first European river bridge to be built using concrete box-girder construction.**

KOCHERTAL VIADUCT, GERMANY (1979).

KYLESKU BRIDGE, SCOTLAND (1984).

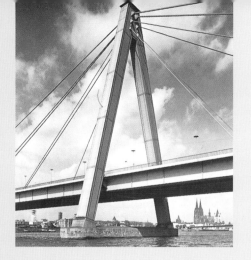

SEVERINS BRIDGE, COLOGNE, GERMANY (1959).

GANTER BRIDGE, SWITZERLAND (1980).

PONT DE NORMANDIE, FRANCE (1994).

ERSKINE BRIDGE, SCOTLAND (1974).

3

What makes a bridge stand up

In understanding how bridges work there are two important structural terms to recognize—tension and compression. All structures—whether a house, a skyscraper, a dome, an arch, or a suspension bridge—are either in tension or compression. If you can understand how tension and compression forces work, then you will begin to appreciate how bridges are able to span.

What are tension and compression, and how can we recognize them? We may not be able to "feel" the forces in a bridge structure, but we can recognize the effects of tension and compression quite easily. For instance, if we pull on a rope or a string we say that we have put the string or rope in "tension." If you push down on a wooden post or hammer a nail into wood, you are putting the post or the nail into "compression." To find out how it "feels" to be in tension, hold the knob of a closed door and pull on it. Your arm is put in tension: it is being stretched. If you want to feel compression, push with your arm stretched against the door. Your arm is now in compression.

In "tension," we are stretching the structure and trying to lengthen it. By pushing down or "compressing" a structure, we are trying to shorten it. The amount of lengthening or shortening cannot be seen by the naked eye.

ABOVE: Closing the main span of Ting Kau Bridge in Hong Kong.

OPPOSITE: Kurushima suspension bridge under construction in Japan.

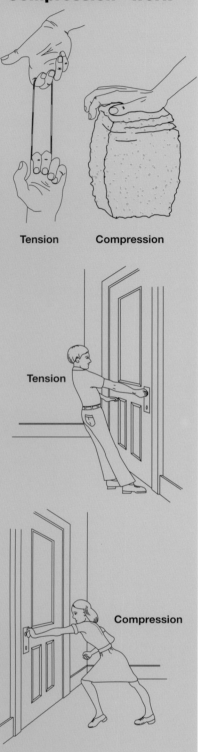

How "tension" and "Compression" work

Tension   Compression

Tension

Tension

Compression

Tension is—pulling a string or rubber band apart or trying to pull open a door that has jammed. **Compression** is—pressing down on a cardboard box, squeezing a sponge or pushing hard against a door to stop it opening.

As a bridge bends under the weight of the load it is supporting, some parts of the structure may go into tension and others go into compression. For instance, an arch goes only into compression, while a suspension cable stretches and undergoes tension—but a truss or beam will undergo both tension and compression at the same time. Materials for bridge building that are useful in compression are stone, brick, and plain concrete. They are brittle and will crack if they are stretched or bent. Materials that are good in tension are rope, bamboo, and wire. They can be stretched, but, because they are not very rigid, they will not hold their shape under compression. However, highly engineered modern materials like steel and reinforced concrete are good in both tension and compression. Successful bridge builders through the ages had to design bridges to suit the building materials that were available to them at the time.

Understanding the tension and compression properties of materials, and how to control these forces in a bridge, helps to determine the type of bridge to be built and the length of its span.

What is Reinforced and Prestressed Concrete? Concrete is an artificial rock comprising sand, stones, and water, which when mixed with cement hardens to full strength in about four weeks. It is produced in ready-mix concrete plants and sent to the bridge works in a concrete wagon or truck, and then pumped into position while it is still fluid and workable. Sometimes it is poured into molds or forms in a casting yard to make precast concrete beams and sections, which are then transported to the site of the bridge works.

When water is added to cement, a chemical reaction takes place, which converts the very fine cement powder into a hard, crystalline rock material. The process is called hydration, and it is a complex chemical reaction. Water is also needed to lubricate the concrete mix to make it workable and easy to pour and compact into molds or forms.

Concrete is very good in compression, but, if you try to bend it or stretch it—putting in into tension—it will crack and fail. For concrete to be ductile, so that it can bend and recover and be capable of resisting tension forces, it has to be reinforced with steel rods or bars—materials called, predictably, "reinforcement." The reinforcement is positioned close to the face of the concrete that is in tension to prevent it cracking and to resist the tension stresses acting on it. Reinforcement placed in concrete will not rust, despite its being covered in wet concrete, because the pH of cement is so alkaline that it inhibits any rust and corrosion forces developing.

Prestressing concrete is a very efficient way of controlling the tension and compression forces within it. Imagine that an axial force is being applied to the concrete—as though we were trying to squeeze it—so that a concrete beam, for example, is put it into precompression before any load is applied. Well that is what is being done in prestressed concrete.

So why is it so useful? By putting the concrete into precompression it will not go into tension when a load is applied, because the precompression stress in the concrete can be made larger than the tension stress from the bending. This is a very simple explanation of how prestressed concrete works. It is much more complex than that in the design of a bridge beam because there are losses in the prestressing strand due to shrinkage of the concrete, and relaxation of the steel stress.

When strands are placed in beams or box girders and tensioned before the concrete is placed in the forms, it is called prestressing. When the strands are tensioned after the concrete has hardened, it is called post-tensioning. In prestressing, the tension wires or strands are cut once the concrete surrounding them has hardened, thus transferring the force from the strand to the concrete. In post-tensioning, the strands are laid in position but are not tensioned.

When the concrete has hardened, anchor plates are placed at the ends of the beam and the strands are tensioned against them, putting the concrete into compression. With prestressed concrete the depth of the beam or box girders is usually 20 percent shallower than the equivalent reinforced-concrete beam, because prestressing is more efficient. It is the preferred method of construction for box-girder beams and bridge decks in modern concrete bridges.

BELOW LEFT: Explaining reinforced concrete.

BELOW RIGHT: How prestressed concrete works.

Plain concrete

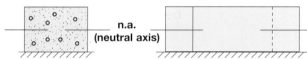

n.a.
(neutral axis)

zero stress

Plain concrete beam resting on the ground has zero stress.

Reinforced concrete

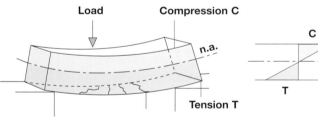

Load   Compression C

n.a.

C

T

Tension T

If a plain concrete beam has to span it will bend— the top half above the neutral axis or the line of zero stress will be compressed, while the bottom half will be stretched and will go into tension.

Bending

Load

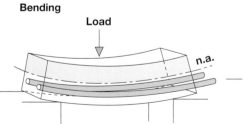

n.a.

Section

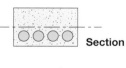

C

n.a.

T

Plain concrete is weak in tension and will crack unless it is strengthened by steel reinforcement to resist the tension stress. (Called a reinforced concrete beam.)

Prestressed concrete

Precompression PC

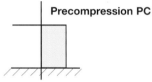

Stress block at rest

A plain concrete beam is prestressed by tensioning steel wire strands that run inside the beam, and which are anchored at each end of the beam. The steel strands compress the beam and impart a pre-compression.

Load

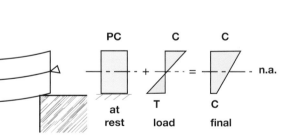

PC   C   C

+   =   n.a.

at   T   C
rest   load   final

When a load is now applied, the prestressed beam does not go into tension when it bends, because the pre-compression stress imparted by the steel strands is much greater than the tension stress. A prestressed beam is more efficient than a reinforced concrete beam.

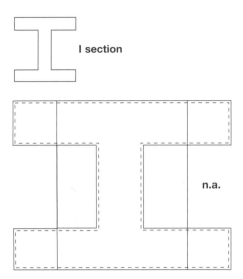

I section

n.a.

The steel beam

Steel is processed and manufactured in factories and then assembled into beam sections, plate girders, trusses, and so forth at steel-fabrication yards, before they are transported to the bridge location. Steel is stronger than concrete and has an elastic modulus that is 15 times greater, making it 15 times more efficient than concrete, but also about that many times more expensive.

To determine the most efficient section and shape for a steel beam, we must consider how it behaves under bending. In bending, the lower part of the beam stretches and goes into tension while the upper part shortens and goes into compression, and in the middle, or neutral, axis it neither stretches nor shortens. This means that the steel at and near the neutral axis does very little work to resist bending, while the steel farthest away from the neutral axis has to resist most of the load. Because of this it is more effective to remove as much of the steel as possible near the neutral axis and place it at the ends. If we start with a beam with a solid rectangular section and then take away the inefficient material near the neutral axis and put it symmetrically about both ends, we get a wide-flange section or I-beam. This is the most efficient shape for a steel beam and, since bending resistance depends on the distance from the neutral axis, the deeper the beam, the greater is the bending resistance of a wide-flange section. The central section of the beam is called the web.

The arch

Early arch bridges The arch bridge is a pure compression structure, and such bridges were in common use at the time of the Roman Empire. Because only brittle materials such as stones and bricks were available, to span any distance more than a large hop they had to employ arch construction to ensure that the forces carried by these materials were always kept in compression.

The Romans built their arches in the shape of a semicircle. It was a simple geometric shape, making it easy to form the centering that supported the wedge-shaped stones, or voussoirs, for the arch. The voussoirs were cut very accurately to the maintain the circular profile of the arch. They were placed symmetrically on the centering, working from both ends of the arch toward the crown.

When the center voussoir, or the keystone, was wedged firmly into place, the centering was taken down. Basically, an arch consists of two halves that lean against each other at the keystone. The weight of the bridge is carried outward along the curving path of the voussoirs. At the

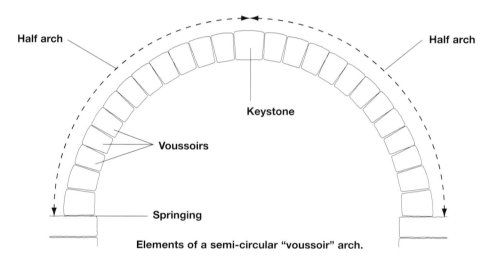

Half arch

Half arch

Keystone

Voussoirs

Springing

Elements of a semi-circular "voussoir" arch.

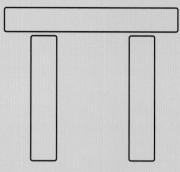

Post and lintel

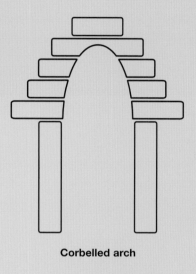

Corbelled arch

Voussoir arch

TOP LEFT: Roman arch construction (Segovia Aqueduct, Spain).

LEFT: Peunta Alcantara over the Tagus in Spain (98 AD).

point where the arch reaches the ground, the downward force from the arch is resisted by piers, while the outward thrust of the arch is resisted by the abutments to prevent it from spreading out and thus collapsing.

For a semicircular arch such as a Roman arch, which met the ground vertically, there is no outward thrust. Thus each arch span can be built independently of the next one, which was useful in Roman times, when they had to build a bridge across a river. However, the pier supports for such arches have to be very wide and massive, and the span quite limited. The flatter the curve of the arch, the greater is the outward thrust on the abutment, and the more unstable the arch becomes during construction.

The two-centered, or pointed, arch is one in which the two halves were formed from the arcs of a circle that meet at a point in the middle. These were built to provide better navigation clearance under the span.

There are different types of arch bridges. The reason for this is that some are more efficient at spanning greater lengths, while others are more economical in the use of materials, and yet others may be faster to assemble.

The segmental arch The segmental arch has a profoundly different structural effect from that of the semicircular arch. The segmental-arch profile is part of an arc of a larger circle. As a result of this the segmental arch introduces sideways or horizontal thrust, because it does not press down wholly in the vertical direction when it meets the ground. The flatter the arch the greater the horizontal thrust.

The advantage of the segmental arch is a flatter arch profile that spans farther and requires less material in constructing the arch and roadway it supports. Requiring fewer piers in the stream or river crossing meant less obstruction to navigation and to seasonal floodwaters. The abutments built at each end of the bridge receive the horizontal thrust from the segmental bridge span.

In a bridge with more than a single span, the horizontal thrusts of adjoining spans are equal and opposite. Thus multiple arches could be built to hold one another up, as long as they were supported by abutments at each end of the bridge. Earlier bridges needed heavy piers to act as abutments in order to build the spans independently. This meant that the piers could be made slimmer because they had to carry only the vertical load, since the arch spans were interdependent.

The stiffened-slab arch This was a discovery of Robert Maillart, who exploited the plastic qualities of reinforced concrete to build very slender, remarkably modern-looking, three-hinged and stiffened-slab-arch bridges at the turn of the century. At each abutment the arch begins as a pair of legs which thicken and fuse toward the quarter-span. At this point the arch joins the bridge deck or road slab.

Over the central section the bridge deck and arch come together to form a stiff box section. The forces in the structure are designed to concentrate at the three points or hinges—one at each abutment and one at the crown. The stiffened-slab arch was characterized by a rigid and heavy bridge deck supported on a thin slender arch slab via pencil-thin, vertical walls. The rigidity of the deck restricts any lateral movement of the bridge because it is joined to the arch at the crown and anchored firmly at each end. It also distributes the load from the bridge deck evenly over the whole bridge, so that buckling in the arch is prevented.

TOP: Example of Maillart's stiffened-slab concrete arch. Schwandbach Bridge, Switzerland.

ABOVE: Tavansa, Switzerland, another stiffened-slab arch bridge designed by Maillart.

Modern arch bridges Modern arch bridges are built mostly of reinforced concrete and steel. The concrete of the arch is poured into wooden or metal forms supported on a scaffolding, once the steel reinforcement is in place. The formwork is removed when the concrete has matured and fully hardened. Usually a concrete-arch bridge consists of two or more parallel arches, from which columns of varying lengths rise to support the bridge deck. The parallel arches are interconnected by struts crisscrossing each other, which makes the arches work together against the lateral pressure of the wind.

Steel-arch bridges are built very similarly to concrete-arch bridges, with a series of wide-flange plate-girder or box-girder beams or tubular-steel elements, which are prefabricated in sections, to the required profile of the arch. These

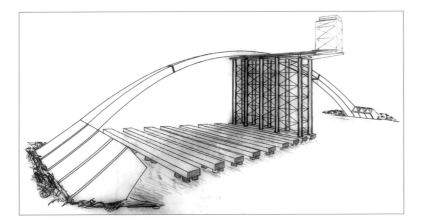

elements are transported to site, and temporarily supported in position, until they are welded and bolted together. Stub columns of steel of varying lengths are built up off the arch to support the bridge deck platform. Steel arches have certain advantages over concrete arches, since they are lighter in dead weight, require smaller foundations and need less supporting scaffolding. They can also be built by cantilevering each half of the arch and connecting them at mid-span.

Modern arch bridges can form slender, flat-arch profiles or very pronounced segmental-arch curves, depending on the length of the span required and the ground conditions. If the ground is very poor on the embankment for instance, but better under the river bed, it may be economic to build a deeply curved segmental arch with less horizontal thrust than a flat-arch bridge.

Variations on the arch shape have been developed to exploit its lightweight construction, span range, navigational clearance, and ability to increase or reduce the horizontal thrust depending on the ground conditions. There is the through arch, sometimes called a "sickle arch," where part of the bridge deck is suspended from the arch and part of it supported by columns from below.

TOP: Illustration showing the scaffold support for the concrete arch span of the Gladesville Bridge.

ABOVE: Gladesville Bridge in Sydney under construction.

RIGHT: Example of a steel flat arch (Pont St Ouen, Paris, France).

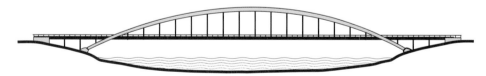

The two hinged arch

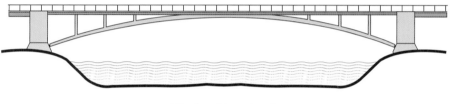

Sequence of flat two hinged arches

Sickle-shaped through arch with suspended deck

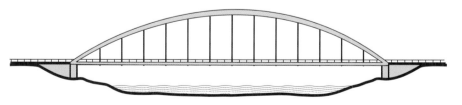

Bowstring, tied arch

ABOVE: Gladesville Bridge, Sydney, Australia.

LEFT: Types of modern steel and concrete arch bridges.

ABOVE: A concrete tied arch bridge under construction (El Rincon Viaduct, Spain).

ABOVE: Constructing the steel arch of the Henry Hudson Bridge, New York.

There is also the "bowstring arch," or tied arch, where the bridge deck is suspended from hangers or truss bracing from the arch. The arch is tied by the span of the deck, which resists the outward thrust of the arch, so that no abutments are required. This type of arch was popular for rail and road bridges across the world, where the river bed was deep or the embankment was quite shallow, making navigation clearance critical.

RIGHT: The completed steel arch of the Henry Hudson Bridge.

Truss girders

One of the most simple and basic of bridge forms is the triangular truss. A simple triangular truss is made up of two inclined compression members and a horizontal tension bar or tie rod, which prevents the inclined members from opening up under load. The truss is self-contained and is supported from each end, to carry the imposed load as well as its own weight. It does not need abutments because there is no horizontal thrust.

In some long-span trusses, the sagging of the tie rod under its own weight is reduced by support at mid-span from a hanger suspended from the top of the truss. Large trusses or composite trusses can be made by connecting triangular trusses together. Highway and railroad bridges spanning hundreds of feet are often built with trusses like this, mostly made of steel bar.

The composite truss behaves very like a deep-flanged beam, with the truss upper chord acting like the upper flange of a steel beam, and the lower chord like the lower flange of a beam, and the inclined or diagonal bracing bars as the web

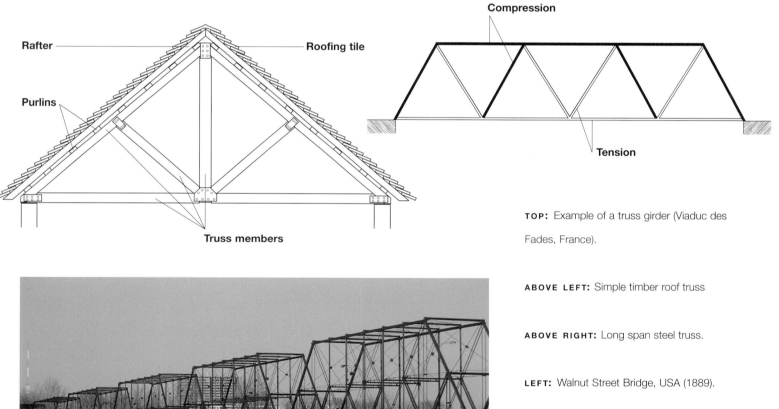

TOP: Example of a truss girder (Viaduc des Fades, France).

ABOVE LEFT: Simple timber roof truss

ABOVE RIGHT: Long span steel truss.

LEFT: Walnut Street Bridge, USA (1889).

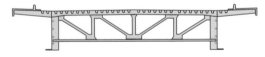

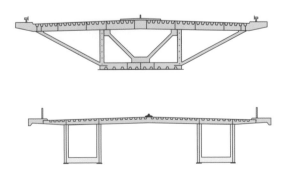

of the steel beam. The diagonals of the truss connecting the upper and lower chord of the truss are much lighter than a full web and use less material. The rigidity and stiffness of the truss is due to the triangulated diagonals.

Gustave Eiffel's great bridges and towers were based on the lightness and rigidity of the truss, because it was very strong and yet economical on materials.

Box-girder and trapezoidal box bridges

Steel beams, plate-girder beams, and reinforced- and prestressed-concrete beams are suitable for relatively short spans for road and railroad bridges, but will require many supporting piers if they are to be built across a wide river or freeway interchange. By forming a box section which is hollow in the middle, a very strong but economical girder beam can be built, which has a greater span range. The box-girder beam, whether it is made of steel or concrete, is very common for modern road or railroad viaducts, and has been used extensively for building elevated freeways along mountain ranges or over valleys in Europe. It was primarily developed for the rigid deck construction required for suspension or cable-stay bridges. The box-girder bridge was first conceived by Robert Stephenson for his Britannia Rail Bridge over the Menai Straits in Wales.

RIGHT: Crane erected, balanced cantilever, cast-in-place box girder construction.

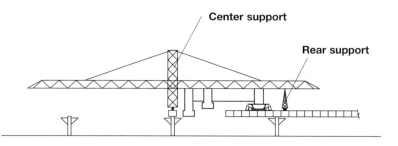

Center support

Rear support

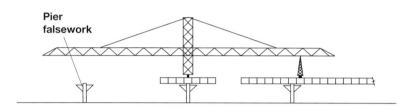

Pier
falsework

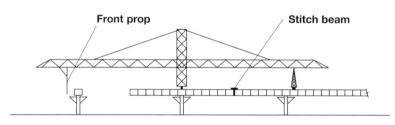

Front prop

Stitch beam

Gantry weight supported by front and rear
supports, center support moved to next pier

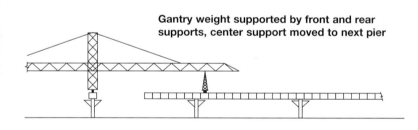

ABOVE: Sequence of box girder cantilever construction.

TOP: Example of a steel box girder with inclined struts.

ABOVE: Cast in place, cantilever box girder construction.

LEFT: Gantry erected precast box girder span.

OPPOSITE: The suspension cable and walkway of the Akashi Kaikyo Bridge, Japan.

OPPOSITE RIGHT: The deck structure is suspended from hangers connected to the main cable (Kurushima Bridge).

FOLLOWING FOUR PAGE SPREAD: "Record Breaking Suspension Span" of the East Bridge of the Great Storebelt Crossing, Denmark.

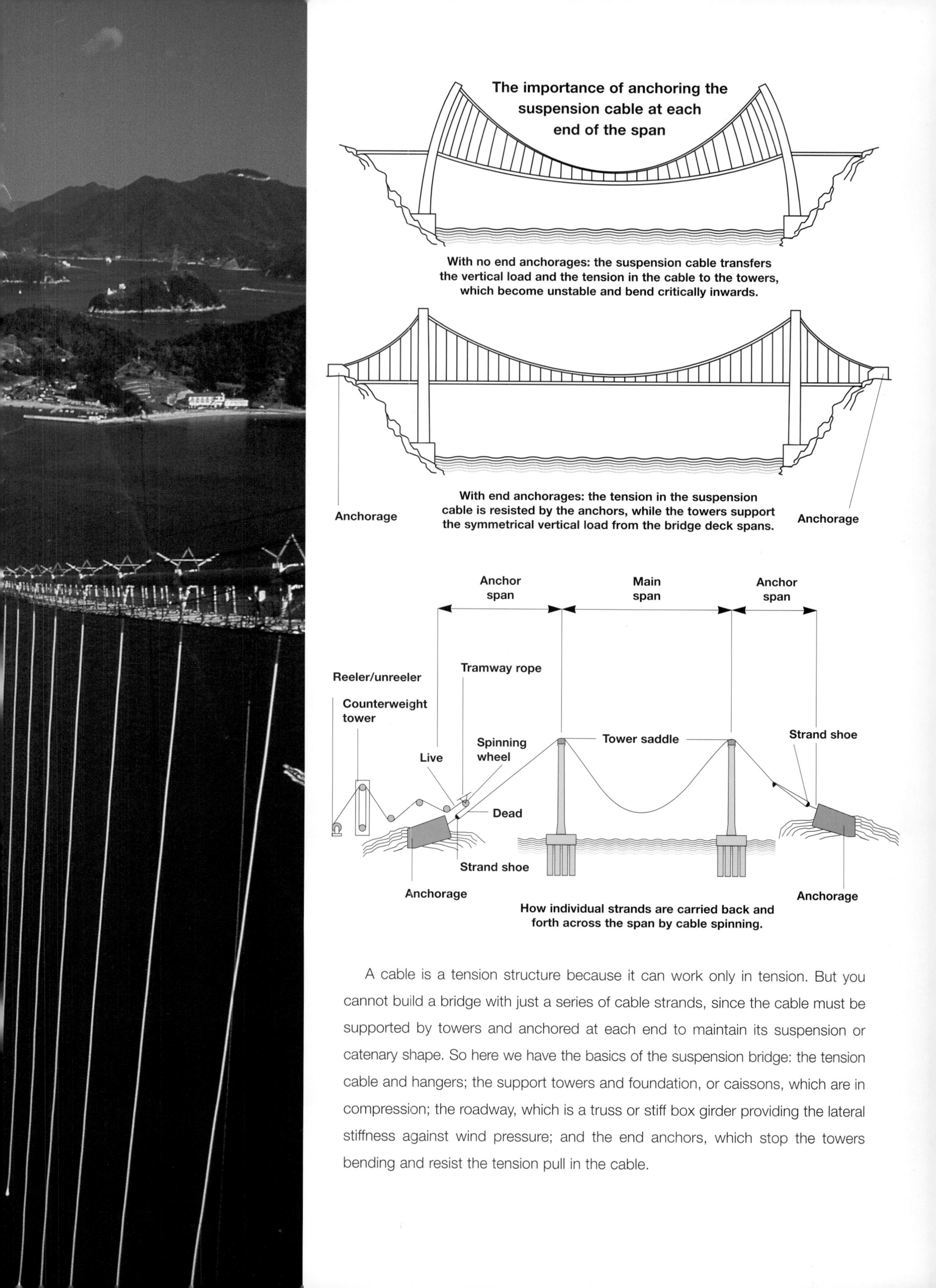

The importance of anchoring the suspension cable at each end of the span

With no end anchorages: the suspension cable transfers the vertical load and the tension in the cable to the towers, which become unstable and bend critically inwards.

Anchorage

With end anchorages: the tension in the suspension cable is resisted by the anchors, while the towers support the symmetrical vertical load from the bridge deck spans.

Anchorage

Anchor span

Main span

Anchor span

Reeler/unreeler

Counterweight tower

Tramway rope

Spinning wheel

Live

Tower saddle

Strand shoe

Dead

Strand shoe

Anchorage

Anchorage

How individual strands are carried back and forth across the span by cable spinning.

A cable is a tension structure because it can work only in tension. But you cannot build a bridge with just a series of cable strands, since the cable must be supported by towers and anchored at each end to maintain its suspension or catenary shape. So here we have the basics of the suspension bridge: the tension cable and hangers; the support towers and foundation, or caissons, which are in compression; the roadway, which is a truss or stiff box girder providing the lateral stiffness against wind pressure; and the end anchors, which stop the towers bending and resist the tension pull in the cable.

Suspension bridges

Natural vegetable fibers like hemp and bamboo have been used for thousands of years to make strong ropes from which to suspend crude bridges or to secure masts and sails in strong winds. Nowadays we make thick steel rope by twisting individual steel wires to make a cable strand, which can measure a few inches to a few feet in diameter. The steel cable is protected by a plastic nylon sheath to prevent corrosion. Such cables can carry thousands of tons of force and are the basic load-carrying system of the longest type of bridge in the world—the suspension bridge.

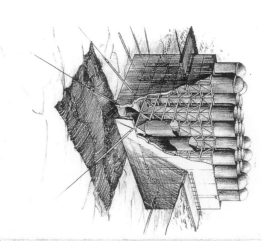

RIGHT: Illustration showing how the caisson foundations and piles for the piers of the Tagus Bridge, Portugal, were constructed.

LEFT: Hanger connections being checked from the main suspension cable.

BELOW: Spinning the individual strands of the suspension cable.

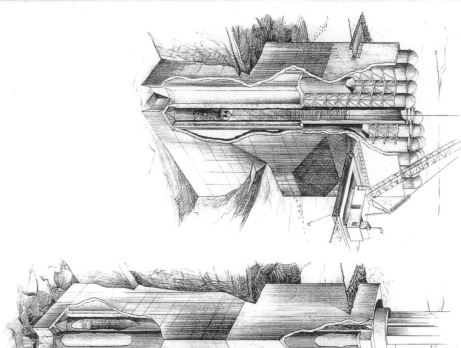

How does the suspension bridge keep its shape as traffic moves over it? The cables, which will want to distort under load, are stabilized by the hangers that support the roadway from the suspension cable. As a load goes over the bridge at a particular point, the cable is held in shape by the hangers and the rigid bridge deck. To keep the horizontal distance between the cables constant, the bridge deck acts like a stiffening truss and maintains the shape of the suspension system.

The larger the sag of the cable between the towers, the less tension is in the cable, which in turn means smaller-diameter and less costly cables and smaller end anchors. The problem is that, in creating a larger sag, the towers that support

ABOVE: Looking through the cable saddle at the top of the tower of the Mackinac Bridge.

LEFT: Floating out the caisson and pile tubes for the Tagus bridge foundations.

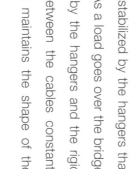

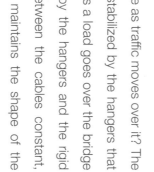

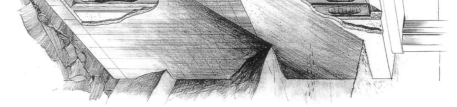

the cables have to be built very high, and this extra cost often outweighs the saving in the cable costs. That is why many suspension bridges have a shallow curvature of their suspension cables. However, the shallow curvature increases the tension in the cables, so thicker cables and large anchor blocks have to be built to resist this force.

The cables of a suspension bridge can weigh hundreds of tons, making it impractical to make the entire cable on dry land. Instead the cables are "air-spun," very like in a textile loom, with individual steel wires pulled across in pairs by a spinning wheel. To set up the spinning wheels, pilot ropes are taken across the span and hauled up over both towers and linked back to the anchors at each end. A catwalk is then assembled from the pilot ropes for each cable, suspended a few feet below the eventual position of the cables.

The spinning guide wires are then sent across and large pulley wheels—one at each end—create a continuous loop for the spinning wheel to track. A section of steel wire is then attached to each of the two spinning wheels which pulls the wires across the entire span. Each time the spinning wheel covers the entire span

it lays two wires fed from two large reels of wire. The spinning wheel runs the continuos loop back and forth across the span until the required number of wires have been positioned.

When the spinning wheel runs over the tower, the wires are placed in special metal saddles at the top, which are carefully located to ensure there is an even spread of the compression force pushing down on the tower from the tension in the cable. The wires are packed tightly together by a special clamping machine, then covered with wrapping wire, painted, and sheathed in a plastic nylon covering for protection. Cable bands are clamped at regular intervals, to attach the hangers that support the bridge deck.

ABOVE: With the anchor blocks and the suspension cable in place, the approach span construction begins on the Mackinac Bridge.

Usually suspension bridges have caisson foundations, which for the East Bridge of the Great Storebelt crossing in Denmark were prefabricated in concrete in a dry dock from where they were floated and then towed out to the center of the estuary. The caissons were ballasted for buoyancy during towing and, once in position, they were flooded to sink down onto the prepared bed on the bottom of the river estuary. The caissons were injected with sand and concrete, displacing all the entrapped water within their voided compartments to form a rigid and stable foundation. The suspension towers are built up from the caisson.

BELOW: The cable strand shoes of the anchor block of the mighty Verazzano Narrows Bridge.

RIGHT: Segments of the truss deck structure are attached to the hangers of the Verazzano Narrows suspension bridge.

BELOW: The East Bridge of the Great Belt crossing in Denmark.

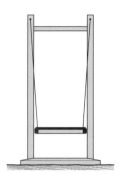

Cable-stay bridges

This type of bridge is a relative newcomer in the bridge world and was developed during and immediately after World War II. They are more efficient than suspension bridges over shorter spans, requiring fewer cables to support the bridge deck. Instead of long draping suspension cables, the bridge deck is supported by a series of individual cables connecting the bridge deck directly to the pylon or cable mast, with the cables arranged in either a fan configuration or a harp configuration. In the fan configuration all the cables run over the top of the mast via a saddle, while in the harp configuration they run parallel to one another and are connected at equal intervals along the bridge deck and the cable mast.

All the early cable-stay bridges used two planes of stays, with cables from twin towers supporting both edges of the bridge deck. The Severins Bridge at Cologne was the first to use a giant A-frame tower to support the fan stays of this asymmetrical-span bridge.

As the trapezoidal box-girder bridge deck was developed for the suspension bridges, to provide an aerodynamic shape to minimize buffeting effects of the

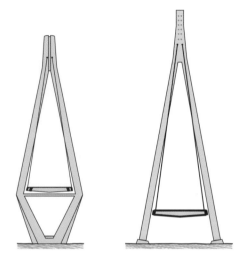

ABOVE: Shape of pylon towers for supporting a "double" cable stay arrangement.

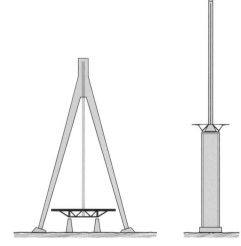

ABOVE: Shape of pylon towers for supporting a "single" cable stay arrangement.

RIGHT: Different types of cable stay bridges.

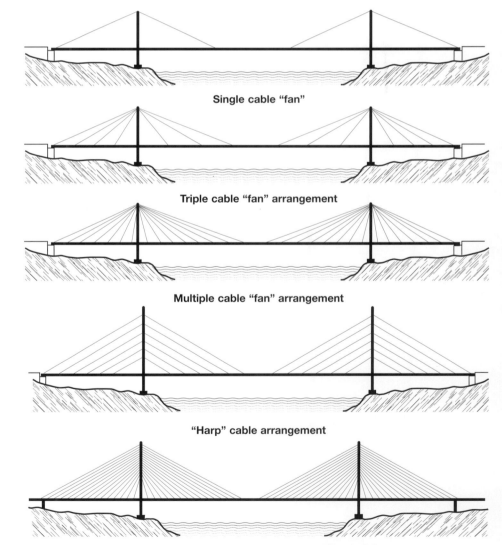

Single cable "fan"

Triple cable "fan" arrangement

Multiple cable "fan" arrangement

"Harp" cable arrangement

Combined "fan" and "harp" cable arrangements

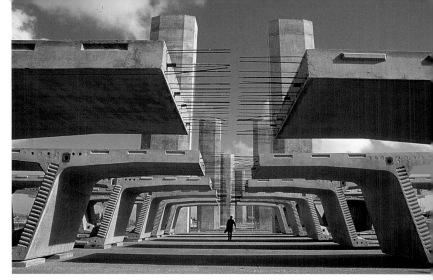

TOP: The pile foundation supports for the approach spans.

ABOVE: Erecting the concrete box sections by crane for the approach spans.

TOP: Precast concrete segments in the casting yard.

ABOVE: Completing the box girder approach span.

wind, they became torsionally stronger and more rigid. This allowed cable stays to support the bridge deck with just a single plane of stays, instead of a pair for each side of the deck. The stays are connected along points within the central reservation of the road or bridge deck. Visually the single plane of stays is more elegant and less fussy than two planes of stays.

For a cable-stay bridge to work properly the cables themselves have to be extremely powerful and must act on the bridge deck with as much rigidity as possible. In the early days several bridges were not successful, since the cables "stretched" too much under cyclic loading. To counter this the modern stay cable is "lock-coiled," whereby the strands in the cable are specially wound to restrict their unwinding under tension.

Constructing the Vasco da Gama Bridge, Portugal

This is an immense bridge project which took many years to plan and design and just 18 months to construct. It represents a world record for a bridge of this scale. The central navigation span is a cable-stay bridge, while the others, called the approach spans, are shorter, concrete, box-girder spans, which have been built using the balanced cantilever method.

Here's a short summary of the construction stages to highlight the main features of the project.

1. Prior to starting any bridge work a marshaling yard, a precast production yard and fabrication site is established. This is where all the materials and equipment for building the bridge are delivered or made—for example, the steel sections, the erection gantry, the precast-concrete box girders, the huge caisson foundations, the scaffolding and temporary works, the cable-strand reels, the cranes, the piling rig, the access platform, the floating pontoons, the dredgers, and so forth. The site can be as large as 74 acres, with offices and parking lots to accommodate over 3,500 people.

   In the case of the Vasco da Gama Bridge, the foundations for the central span's piers were the first to be completed. They consist of 44 bored piles which were 165 feet long and just over 7 feet in diameter, surmounted by the pylon base. The base is a concrete raft 275 feet by 65 feet in plan and about 10 foot thick, and has prestressed beams, which connect the two legs of the H-shaped pylon to the base. The pylon base ensures the distribution of the vertical forces from the bridge and resists the horizontal thrust which may arise owing to impact from shipping.

2. The pylon legs were then built up from the base, by a climbing formwork system. The reinforcement for the pylon legs was placed in the 13-foot-high formwork section, before the forms were clamped. Concrete was poured into place, and the forms were eased away once the concrete had hardened. The forms were crane-lifted to the next 13-foot-high section, and adjusted to the

correct width and taper of the pylon profile, before being concreted again. Stage by stage, the climbing formwork rose to create the legs of the 490-foot-high shafts of the twin pylon supports.

3. Next the pylon crossbeam was cast, before the cable support section of the twin pylon masts was built. The top section of the pylons was formed using precast-concrete box elements that stacked one on top of another, all the way to the top. The box sections were precast on shore and transported by barge to the bridge site. Each section was lifted onto the previous section by a barge-mounted crane, and then jointed using *in-situ* concrete. Steel sections were cast into box sections to provide the anchorage for the cable stays.

4. The prefabricated steel formwork to cast the bridge deck was then assembled at the base of each pier. The pier section of bridge girder deck was cast and the 2,000-ton assembly was then jacked into position. As the bridge is in a

BELOW LEFT: Building out the central span.

BELOW RIGHT: Connecting the cable stays to the bridge deck.

seismic zone and has to withstand earthquake movements of up to 4 feet 3 inches across the width of the span and 6 feet along it. In order to withstand this movement the bridge deck is suspended directly from the pylon and not connected to the crossbeam on the pylon masts, which is more usual. Damping devices diminish the deck movement and ensure the bridge-deck-to-pylon linkage.

5. Once the pier section of the bridge deck was in position, the steel formwork gantry assembly was cantilevered out from both ends of the pier section, to allow 29-foot sections of the bridge deck to be concreted. The deck assembly was held in position by the permanent cable stays, which were connected to the pylon mast. Slowly, in 29-foot lengths, the bridge deck started to span across the 1,380-foot central gap.

   Built as an extension of the main span, the central viaduct has a length of 21,370 feet and is composed of 81 bays in total, 75 of which span 258 feet, two span 426 feet over a navigation channel, and four span 308 feet. In cross-section the concrete box-girder deck consists of two parallel girders, one for each carriageway, and has a standard depth of 13 feet, which increases to 26 feet for the longer spans, with the top section of the box girder forming the roadway slab. The box girders are haunched (get deeper) over the pier supports, creating a flat-arch profile.

6. The concrete box sections were cast 15 miles upstream of the bridge. Each section was between 23 and 33 feet long, weighing around 256 tons. Cast on an assembly line, the box sections were connected in groups of eight to form a prestressed beam 256 feet long and weighing 2,165 tons. They were towed out in pairs and lifted into place by barge-mounted cranes, and then stitched onto the central pier section using prestress, to maintain the balanced cantilevers. The southern approach viaduct has an overall length of 12,550 feet and was built using gantry-launched, balanced cantilever construction. The four deep beams for each span were cast in place, using movable formwork.

On March 31, 1998 the Vasco Da Gama Bridge was officially opened.

TOP: Nearly two-thirds of the cantilevered half span is now complete.

ABOVE: The view looking along the centerline of the finished cable stay spans.

LEFT: Aerial view of the distinctive cable stay spans of the Vasco da Gama Bridge over the Tagus estuary in Lisbon, Portugal (1998).

4

Great bridges—the bridges of destiny

It is never easy to make a definitive list of the greatest bridges of the world. In making a short list, it seems logical to select bridges that are representative of the great periods of bridge building over the centuries, such as the Roman, Medieval, Renaissance, the Age of Iron, the Industrial Age, and so on.

At the same time we are also drawn toward a number of spectacular bridges that were built during this century; but these have been depicted in every book on bridges in the past 20 years.

Taking my cue from the great bridge-building eras, I have made my selection with the proviso that the bridge must still be standing today, it must be accessible to the public, and must sit in a wonderful location. It was also important that the images available should be stunning and, for the contemporary subjects, images of the bridge taken both during construction and after completion should be shown, to contrast that with pictures taken more recently, if that was possible. It is important to highlight the working conditions during construction—the dangers and the risks—and the plant and equipment that were used to build the bridge. It gives a sense of the scale and hardship that were endured during construction, which may never be fully appreciated when viewing the finished bridge.

ABOVE: Building the cantilever steel arch of the St Louis bridge over the Mississippi, USA.

OPPOSITE: The Sunshine Skyway Bridge in Florida hit by lightening during a thunderstorm.

Pons Augustus, Rimini, Italy (*c.* AD 14)

Many consider this bridge over the Marecchia at Rimini the finest bridge of the ancient world and greatest bridge ever built by the Romans. It was built during the reign of Augustus (31 BC—AD 14). The genius of Roman bridge engineers seen in the Pons Augustus, the Puenta Alcantara and Segovia Aqueduct in Spain, Trajan's Bridge over the Danube, and the Pont du Gard in Nîmes expresses the power and might of the Roman Empire better than all the ruins of the Forum in Rome. Palladio in the sixteenth century declared Pons Augustus the finest bridge in the ancient world and copied the spans many times in his own construction of bridges. And, because Palladio's bridge designs were much admired, they were also copied by bridge builders in later centuries. We can see second- and third-generation "Pons Augustus" in many European cities today.

The Pons Augustus has five spans, the three middle ones are 28 feet, and the two end ones are 23 feet. The whole structure is unifying. On each spandrel above the piers, there is a panel framed by pilasters that uphold a classic pediment. The balustrade is solid, its heavy cornice supported on toothlike stone supports called dentils, while the face of the bridge was covered in marble. It was unusual for a bridge built outside Rome to feature such charming decoration.

From a structural viewpoint the Pons Augustus is interesting because it is the earliest known example of a bridge built on a skew. The piers are not at right angles with the axis of the bridge, although the amount of skew is small. It is probable that the bridge builders did this to locate the pier foundation in the river bed, where the soil and the current made construction easier.

When the construction methods of the Romans are recalled, the Pons Augustus, along with all the other great bridges, will be remembered with awe. Generally, slaves did the laboring, hauling the stones, lashing and nailing the timber centering, building cofferdams in fast-flowing streams, under Roman supervision. Crushing and drowning must have been frequent. They used the simplest of tools—the wedge, the lever, the pulley, and the inclined plane. For Trajan's Bridge over the Danube at Turnu Severin in Romania—the longest bridge ever built by the Romans at 3,000 feet—it is estimated that slaves cut and hauled no fewer than one million rocks of about 2 feet cube, and dropped them in the Danube for the foundations. The time taken to complete a bridge was not a governing factor.

Foundations were built by simply throwing loose stones into the river until they were piled high above the water line or by using open cofferdams and timber piles. The Romans later evolved the art of pile-driving for forming bearing piles in the river bed and cofferdams, which were then enclosed by a foundation bed of concrete, to support the stonework of the piers.

All Roman arches were semicircular and were self-supporting, so that, if the enemy destroyed one arch, the others would remain standing. They generally built the piers to be one-third of the span, and that is why they were so massive.

The Romans knew how to build only the semicircular stone arch, but their mastery of this art has created the most noble structures ever seen or built in the history of civilization.

ABOVE: Pons Augustus as it looks today.

OPPOSITE: Pons Augustus, Rimini as it looked about fifty years ago.

Valentré, Cahors, France (1347) Pont Valentré is a fine example of a medieval fortified bridge. The bridge was built during the period when feudalism was fading out of the medieval world. This was the end of the era of the "Brothers of the Bridge," the Benedictine order of monks we met in Chapter 1. Merchants and public authorities were beginning to turn to local masons and craftsmen to build their town bridges, rather than wait many years for the Benedictine Order to build it for them.

Perhaps this might explain why it took 39 years to complete Valentré. The people of Cahors wanted a new bridge over the Lot, but, not wishing to break with the church entirely, they appointed the Bishop of Cahors as the nominal head of the project. The townspeople collected the money for the bridge by imposing a duty on all merchandising and commodities entering the city gates. It took from 1308 to 1347 to collect the money and to complete the bridge. Slowly and surely Pont Valentré was built, never more than two arches at time, so that building work could be easily adjourned without problems if the work was interrupted by war. (War was also the reason why medieval bridge builders made arches semicircular and stable and each pier an abutment, because if one arch was destroyed it would not cause the others to collapse.)

In regarding the bridge, the eye is drawn to the three towers rising 130 feet above the river Lot. The towers have tiled roofs covering their crenellated walls, from which fire could be directed at boats in the river as well as the approaching troops on the bridge. The six two-centered, arch spans have a width of 45 feet and have a recessed arch ring which accentuates the voussoir stones. The piers are 18 to 20 feet wide with triangular cutwaters, on both the upstream and downstream faces, that rise up to the roadway.

A series of holes can be found in the piers just below the springing line of the arch. Through these holes workmen pushed fir saplings until they jutted out at the either side of the pier. These supported a planked floor which provided a walkway for the workman, a resting place for the stone blocks, and a foundation for the centering. It is not usual to find piers so large in medieval bridges because the builder, being unable to calculate the actual stresses and load carried by the piers, used empirical rules to make sure they were safe.

Valentré also comes with a legend of the "Devil's Bridge," which was probably propaganda spread by the church about bridges that were not built by the Brothers of the Bridge. The legend states that the builder of Valentré sold his soul to the Devil after being discouraged by the slow progress made in the building. But the builder made one proviso: he would give his soul provided the Devil performed every task he was given by the builder. Work progressed very well until the builder had second thoughts about parting with his soul.

He thought he would catch the Devil out by asking the him to carry water in a sieve for the masons. No matter how fast the Devil flew, all the water had gone

BELOW: The medieval Pont Valentré over the river Lot in Cahors, France.

from the sieve by the time he reached the masons. The Devil admitted defeat and released the builder from his spell, but left him with this thought: "You have won," said the Devil, "but I will wager you that you do not boast of having had my gratuitous collaboration!"

When the bridge was completed and the builder was completing the central tower he found that a stone was missing in the northwest corner of the roof. He instructed the workmen to replace it and this they did, but, when they began work the following morning, it was missing. This went on day after day, with the builder replacing the missing stone only to find it missing the following day. The builder grew weary of the Devil's games and to this day—according to the legend—this stone is still missing.

Santa Trinità, Florence, Italy (circa 1570) Scholars and writers on arch bridges are unanimous in their vote for the best-engineered bridge structure of the Renaissance. For purity of form and its pencil-slim elliptical arches, the Santa Trinità over the Arno in Florence is without equal. The curve of the two arch spans defies analysis and has mystified engineers over the centuries when they have tried to explain its nature and how it was achieved. The arch consists of two curves, each resembling the upper part of a parabola, which meet at the crown at an obtuse angle, and the point at which they meet is concealed by a decorative cartouche or pendant. How did bridge builders in the sixteenth century know how to construct such a slender arch span, with the limited technology and crude mathematics that they knew?

RIGHT: The famous statues sculptured by Francaville, restored to the Santa Trinità Bridge after they were destroyed in the second world war.

The curve is very shallow, and has a gradient of one in seven, which is startling when you consider that most arches built around this period had gradients of one in four at the most! It is also a breathtakingly elegant bridge.

In 1567 Grand Duke Cosimo I ordered a masonry arch bridge to replace the old wooden Santa Trinità over the Arno. Bartolommeo Ammannati was the Grand Duke's chief engineer and was asked to design and supervise the building. All sorts of theories abound as to how Ammannati arrived at the curve of the arch. There are no records to provide the answer, and, since there were no mathematical procedures laid down at the time, there seems to be only one answer. Ammannati designed it from aesthetic judgment, being a good draftsman and artist, although it is thought that Michelangelo had a hand in the profile.

Ammannati apparently drew two tangents as determinants. One was drawn vertical to the springing and the other slightly inclined to the horizontal at the crown of the arch. These tangents were then connected by a graceful curve, one for each half of the span. These meet at a slight angle at the crown, which is why Ammannati hid it with the cartouche. The Santa Trinità stood undisturbed for over 300 years, until 1944, when it was demolished by the Germans in World War II. It has been lovingly and painstakingly rebuilt and restored to its original glory.

The four famous statues that represent the four seasons, and which were added to the bridge in the seventeenth century, by the sculptor Francaville, have also been recovered.

ABOVE: The Santa Trinità Bridge over the Arno in Florence, Italy.

Rialto, Venice, Italy (1591)

"The best building raised in the time of the Grotesque Renaissance, very noble in its simplicity, in its proportions, and in its masonry." That is what John Ruskin wrote to describe the Rialto Bridge in Venice. For the first time in the history of bridge building, bridges were designed as architecture that was befitting the civic spirit of their community, to be graceful and aesthetically pleasing as well as being functional and well engineered.

The designer of the bridge was Antonio Da Ponte, a surname that means "bridge." It was a common name in Renaissance Italy. Not much is known about the hero of this story, who was a native of Venice and a curator of public works. The most remarkable fact about Da Ponte was that he was 75 when he won the contract to design and build the Rialto. His fame as a designer came to the notice of the Venetian Senate when he saved the Ducal Palace from complete destruction, after a serious fire broke out in the Palace in 1577. In spite of the dangers of collapsing timbers and molten lead spilling down from the burning roof, Da Ponte entered the flaming building and took charge of the situation. It was his courage and presence of mind that saved the palace. He also argued convincingly against Palladio's suggestions that the palace should be demolished and a new building erected after the fire.

Da Ponte saved for posterity one of the best examples of Italian Renaissance architecture at the same time winning for himself the distinction of designing all the repairs. When in 1587 the Senate invited designs for the Railto, Da Ponte won the contract, beating designs by Palladio, Michelangelo, Fra Gioconda, and other such eminent architects.

BELOW: Rialto Bridge, Venice, Italy.

LEFT: The central gallery and walkway of the Rialto Bridge.

BELOW: The central arch over the covered walkway.

The bridge stands today just as Da Ponte had designed it. In order to give free access to the canal boats, a single segmental arch covers the distance between the abutments. The arch has a clear span of 88 feet, and rises 20 feet above the springing. The total width of the bridge is 75 feet and it has a roadway down the middle, with shops on both sides and a footpath and parapet down each side.

Two sets of arches, six on each side of the large central arch, support the roof and enclose the 24 shops within it. It took three and a half years to build and kept all the stonemasons in the city fully occupied for two of them. According to the records 800 larchwood timbers were used to erect the falsework for the arch.

The foundation of the bridge created controversy, although it proved to be of very durable construction. Da Ponte used timber-bearing piles driven into the soft, silty canal bed by a mechanical hammer operated by four men, to form tightly packed rafts. The piled rafts were driven deeper into the canal bed the closer they approached the water. The tops were then trimmed off to form steps going down toward the water. The stepped rafts were timbered and surmounted with a brick platform built at the correct angle to receive the radiating courses of the abutment and spandrel masonry.

This was such a novel method of construction that many well respected engineers in the city raised doubts about the adequacy of the construction, causing the Senate to order the work to stop, pending an official inquiry. After expert witnesses were called in and Da Ponte had explained his method of construction, the Senate decided that the work was safe. The huge cut stones of the arch were now put in place supported on timber falsework built across the canal.

The bridge was opened in July 1591 and has stood the test of time remarkably well, even withstanding an earthquake without a crack, when Venice was shaken by a tremor shortly after the Rialto opened.

Pont de la Concorde, Paris, France (*c.* 1796) The greatest stone-arch bridge designs of the eighteenth century came from the inventive genius of Jean-Rodolphe Perronet, but, alas, his finest bridge, the Pont-Sainte-Maxence over the Oise, is no longer standing, having been destroyed in World War II. However, his most celebrated bridge, and the last bridge that Perronet designed, is still standing today. Perronet's discovery of the continuous action of thrust in segmental-arch construction, and his invention of the divided pier, completely

ABOVE: Building the arch supports of Pont de la Concorde over the Seine in Paris with timber centering. The water wheel, seen in the background, drained the coffer dam.

changed bridge design for all time. His arch bridges were more slender and spanned farther than any other. Perronet wished to make the Pont de la Concorde in Paris his last work, the most remarkable bridge structure ever built, with arches that were nearly flat and piers that were composed of just two Doric columns. But his vision was compromised by the ignorance of meddling officials and engineers who did not grasp the fundamentals of the design that Perronet had calculated and presented to them. He was 75 when he was invited to design the bridge and, because of his brilliance, there was no one with the intellect to appreciate his ideas. He was forced to increase the arch curvatures and to thicken the piers to satisfy the skeptics. The final result was a less slender bridge than Sainte-Maxience, but Perronet skillfully heightened the appearance of slenderness by introducing a balustrade instead of a solid wall for the roadway, which was the tradition for bridges in those days.

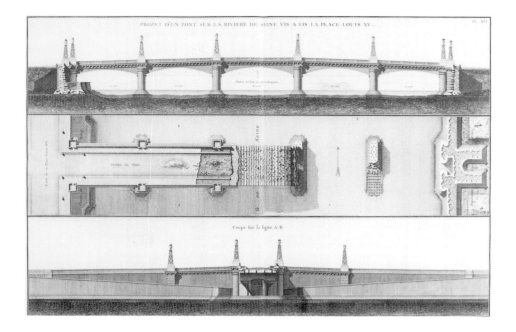

PROJET D'UN PONT SUR LA RIVIERE DE SEINE VIS A VIS LA PLACE LOUIS XV

LEFT: Peronnet's drawings of the Pont de la Concorde.

BELOW: The Pont de la Concorde as it looks today.

Perronet was 80 when work started on the foundations of the bridge. They were of the usual design of wooden platforms on timber piles, about 6 feet 6 inches below water level. When the bridge was half built, the French Revolution broke out. The chaos that followed did not affect work on the bridge. Perronet oversaw everything and lived in a little house built at one end of the bridge. And when the Bastille was razed to the ground the stones provided a good source of masonry! When crossing the Pont de la Concorde, you are walking on history … the ghost of the legendary "man in the iron mask" probably walks there today … Many of the laborers who built the bridge belonged to the military force of the community of Paris, the formidable force that beat the National Guard and tore down the Tuileries.

The bridge has a noble and monumental presence. Its graceful lines and skillful construction make it an artistic triumph. It is clad in marble dressed with a classical ornamentation along its flanks. The decorative feature of the corbeled cornice is believed to have been inspired by the Pons Augustus in Rimini.

LEFT: The balustrade of the bridge is so reminiscent of the balustrade of the ancient Pons Augustus.

St Louis Bridge, St Louis, USA (1874) While Europe continued to build with wrought iron until the 1900s, American bridge designers took the bold decision to experiment with steel and to produce the first major bridges using it. The St Louis and Brooklyn bridges were the first bridges in the world to use steel as the principal structural material. Before this, steel was used only with wrought iron, as an expensive but efficient material for longer-span bridges. Steel was used for the main arch ribs on the St Louis Bridge, and for the cable stays and bridge deck trusses on the Brooklyn Bridge. Both bridges were under construction at the same time, but the St Louis was completed nine years before the Brooklyn.

James Buchanan Eades was one of the most gifted engineers of America in his day. You could say he was a latter-day Leonardo da Vinci, setting up many new ventures, and establishing many successful marine and industrial companies in St Louis during his lifetime. And, when he was approached to design a bridge over the mighty Mississippi, he designed the most daring structure ever seen.

Eades had never built a bridge before, so why did the city fathers and bridge sponsors trust him with the task? For 20 miles above St Louis, where the Missouri empties into the upper Mississippi, it turns into a deep and seething torrent, which is subject to great changes, both seasonal and tidal. The speed of the river can vary from 4 feet per second to 12 feet 6 inches per second in flood, and has a tidal range of 41 feet from low to high tide. It's a brute of a river because, as the river flow increases, sand and mud deposited on the river bed is churned up and carried downstream; as it slows, sediment is deposited on the river bed.

The movement of the river bed was understood by Eades better than any other engineer in St Louis, through his extensive studies and personal observations as a riverboat captain, as a designer of iron-clad gunships during the American Civil War, and as a salvage operator on the Mississippi.

Even when the Missouri legislature agreed to grant a charter in 1864 to a consortium calling themselves the "St Louis and Illinois Bridge Company" led by Norman Cutter, there were delaying tactics by rival factions and objectors to the bridge, principally from the ferry boat companies. Meanwhile, more winters were to pass, with the Mississippi freezing over and ice floes scouring the river bed and presenting a hazard to shipping. At times the river was too treacherous for navigation, cutting off communication and traffic for weeks between the east and west of the city.

BELOW: The bridge deck being built up from the arch.

Finally, on August 20, 1867, actual construction was started, but even when the corner stone of the west abutment was laid in February in 1868 a rival bridge group confusingly calling themselves the "Illinois and St Louis Bridge Company" tried to hijack the bridge project from Eades and his backers. The dispute was settled in March, 1868, when both bridge companies merged and Eades was made chief engineer.

In trying to build a cofferdam for the western abutment, the construction team had to drive through 60 years of scrap dumped in the river. Eades had known that there were at least three burned-out steamer hulls, the wreckage of four barges, anchors, chains, and all sorts lying above the bedrock he wanted to reach. All this made it extremely difficult to make a watertight cofferdam. The pieces that could be hauled out of the water easily were removed, but much of the stuff was buried deep under the silt and mud. He designed a gigantic wooden chisel with a steel blade to cut and smash a way through this underwater scrap yard. Eades wanted to found the cofferdam on bedrock, knowing it would be folly to sit the foundation on the gravel layer above it, which could be scoured by the river in flood.

Having driven the west abutment and built up the masonry structure within the cofferdam, Eades now faced the harder challenge. The east abutment was some 130 feet down to bedrock— 60 feet of water and 70 feet of sand and silt—and was deeper than any caisson had been sunk. He studied the reports of the ingenious pneumatic air caisson developed by Brunel at Saltash for the piers of the Royal Albert Bridge and crossed to France to discuss pneumatic-caisson construction with Monsieur Audernt, who was sinking one for a bridge over the Allier. On returning to St Louis he postponed the east abutment until he had built the two-pier foundations in the river, using this technique, as they were shallower.

The caisson was a huge rectangular box made of wood and sheeted in iron panels and stiffened with girder plate. Inside the caisson there were stairwells leading down to the working chamber, a sealed compartment, which was 9 feet high and bottomless. Other airtight shafts were built for removing excavated material out of the working chambers. As the masonry pier within the cofferdam was built up from the roof of the working chamber, the caisson would start to sink into the water until it reached the river bed. To stop water filling the working chamber, air was pumped in to equalize the water pressure.

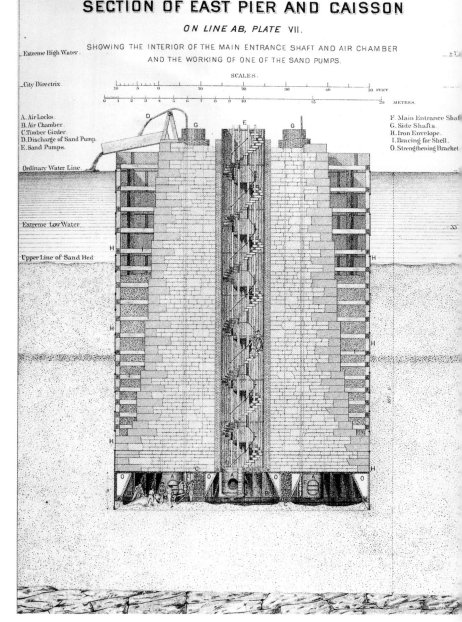

SECTION OF EAST PIER AND CAISSON
ON LINE AB, PLATE VII.

SHOWING THE INTERIOR OF THE MAIN ENTRANCE SHAFT AND AIR CHAMBER
AND THE WORKING OF ONE OF THE SAND PUMPS.

SCALES.

Extreme High Water.

City Directrix

A. Air Locks.
B. Air Chamber.
C. Timber Girder.
D. Discharge of Sand Pump.
E. Sand Pumps.

Ordinary Water Line.

Extreme Low Water.

Upper Line of Sand Bed.

F. Main Entrance Shaft
G. Side Shafts.
H. Iron Envelope.
I. Bracing for Shell.
O. Strengthening Bracket.

ABOVE: Diagram showing the caisson of the east pier.

Once the caisson reached the bottom of the river bed, down the ladders and through the airlock chambers went the workmen. The air pressure inside the chamber was increased as the caisson was excavated slowly down to the bedrock. This could take many, many months depending on the difficulty of alluvium to be removed. Eades had devised a sand pump, the first of its kind in the world, to dispose of the river excavation, rather than use the traditional system of buckets on a rope, which was much slower. Unfortunately, not much was known about decompression sickness or, as it is sometimes called, caisson disease or the bends, caused by working under high pressure and decompressing too quickly. On the east and west pier caisson, there were 91 cases of the bends, with 13 deaths and two persons crippled for life. Eades sought the advice of the best medical brains, and a Dr Jaminet was hired as medical adviser. Jaminet suggested a slow decompression and cutting down the working shift to two hours at 32 pounds pressure; and to one-hour shifts when under 34.5 pounds pressure. There were few problems when this procedure was adopted, although the decompression rate used was still too fast at 6 pounds per minute. The safe figure is 1 pound per minute, but this was not discovered until many years and many bridges later! For the east abutment, which was much deeper, there were surprisingly few injuries, after the working restrictions imposed by the doctor were followed.

Eades next had to present to his promoters and critics the economic and structural justification for building a three-span, steel-arch bridge on a scale never seen before. In his defense of the feasibility of the steel arch, this is what he wrote:

"In 1801 the great Scottish engineer, Thomas Telford proposed to replace the Old London Bridge with one of cast iron, having a span arch of 600 feet … For forty years this remarkable man continued to enrich Scotland and England with some of the most stupendous and successful triumphs of engineering skill to be found in Great Britain … Surely the recorded judgment of such a man as Telford when sustained by the most eminent men of his day, asserting the practicality of a cast iron arch of 600 feet span in 1801, furnishes some "engineering precedent" to justify a span of 100 feet less in 1867."

BELOW: The St Louis Bridge spans the mighty Mississippi (1874).

The fabrication of the steelwork was awarded to the Keystone Bridge Company, one of the finest bridge companies in the USA. Eades had devised test apparatus to assess the structural quality of the steel plate that was rolled to form the tubes of the main arch. After teething difficulties in achieving the grade and strength of steel required, scaled-down sections of the arch were made and tested. These showed that the steel tube for the arches would be more than adequate to resist the stresses imposed on them on the actual bridge.

The three great arches were erected without falsework, built by cantilevering the arches out from the piers toward the center of the span. Temporary towers were erected on the piers to support the tie-backs. The tie-backs or cantilever cables were made of steel bar an inch thick and 6 inches wide. As the arches cantilevered farther and farther out over the water the next problem arose. How to ensure that the two halves meet in mid-span? To ensure this happened Eades built each section bigger by a small factor, which would produce an overlap at mid-span. And if this did not work he designed closure tubes to slide over both sections at the crown.

On September 14, 1873, on the first attempt to close the inner ribs of the western span, there was a gap so small that the closing tubes could not be fitted. A loan of half a million dollars depended on closing the first span by September 19, five days' time. This money was vital to keep the project running. By cooling the steel it was thought it might be possible to close the gap, so the arch rings were packed with ice, which was placed in wooden troughs built around them. Fifteen tons of ice was placed in the troughs on September 15, and by sunrise the following day the gap was still five-eighths of an inch. Forty more tons of ice was placed the next day, but by sundown the arch had closed no further.

With no prospect of cooling weather in the next few days, the ice poultice was abandoned and a special adjustable closure tube that had been designed by Eades was hurriedly fabricated. On September 18, within a day of the deadline, the arch was closed. By December the inner ribs of the central and eastern arches' spans were also closed. It was fairly straightforward now to finish the outer ribs and to build the roadway and railroad deck structure.

On July 1, 1874, Eades organized a public show of the strength and integrity of the bridge, halting 14 heavy steam locomotives—seven on each track—over each span and then sending all 14 in single file over all the three spans.

And on July 4, 1874, Independence Day, the city of St Louis celebrated the opening of the bridge in magnificent style. A mammoth procession paraded through the streets of the city, crossing and recrossing the bridge. In the evening a fantastic fireworks display launched from the roadway of the bridge lit up the night sky and the triumphal arch constructed near the bridge portal that was topped with a portrait of James Buchanan Eades bearing the inscription "The Mississippi, discovered by Marquette, 1673; spanned by Captain Eades, 1874."

ABOVE: The truss section of the railway platform below the roadway deck of the St Louis Bridge.

Garabit Viaduct, St Flour, France (1884) In France the spread of the railroads and the growth of industry were slower than in the United States and Britain. To some extent this was attributable to the remoteness of the raw materials from the towns and cities. Many important mineral deposits were located in the high, barren plains of the Massif Central, which meant opening a railroad before minerals could be commercially exploited and sent to the large towns of Lyon and Limoges. Alexander Gustav Eiffel had been working on the rail network for many years and had built many fine iron viaducts across deep gorges. He had developed methods of designing pylons, towers, and truss girders, to withstand the high winds that funneled down the valleys. Eiffel's designs were capable of resisting the wind forces, and the vibrations that were caused by it. He made sophisticated measurements of the wind force and wind direction and then studied the effects this would have on the structure. The stiff lattice trusses that he designed, with their large open spaces and minimum wind resistance, were Eiffel's solution to building high-masted structures and long-span bridges across steep valleys.

The Garabit Viaduct over the Truyère at St Flour is an important structure because it marks the changeover from iron to steel in bridge construction. The two-hinged-arch concept that Eiffel designed for Garabit was to become the standard design of steel arches that were to follow.

But Garabit is not a steel structure: it is built from wrought iron. There were difficulties in the manufacturing quality of steel at the time, and it was more expensive, so it was not the best material to build with. Eiffel's 530-foot, wrought-iron, parabolic arch supports the 1,850-foot-long truss of the railroad, some 400 feet above the river Truyère. The arch is narrow and deep at the crown to carry the railroad trusses, while at the supports it is wide and shallow. The ends of the arch rest on hinges, which allow for the expansion and contraction of the bridge.

Eiffel pioneered the cantilever construction method for his metal-arch bridges to eliminate the high cost of building falsework below the bridge. The half-sections of the Garabit arch were built out from each abutment and held in position temporarily, using wire stays. Garabit was the longest and highest arch bridge in the world when it was built, exceeding Eiffel's arch truss over the Douro in Oporto by 50 feet. It was finished in 1884 and was the last bridge that he built.

ABOVE: A support pier and the truss of the railway deck.

RIGHT: The changing profile of the arch is clearly visible—from wide and thin at the base to deep and narrow at the crown.

FOLLOWING SPREAD: Close up of the open structured bracing of Gustav Eiffel's wind resistant arch.

George Washington Bridge, New York, USA (1931)

Like the Empire State Building, the George Washington has come to be regarded as one of America's greatest constructions. Crossing the Hudson at its widest point, with a clear span of 3,500 feet and steel towers rising high above the water level, the George Washington Bridge smashed the longest span record by 1,700 feet—a margin that is as much as the actual span of the Firth of Forth! With massive spans like the George Washington, the suspension bridge displaced the cantilever-truss bridges as the accepted type of long-span bridge of the future.

In the George Washington Bridge we see the culmination of the many advances in suspension-bridge construction since the Brooklyn Bridge was finished in 1888. It did have flexible steel towers, which were first pioneered on the wonderful Manhattan Bridge and made it possible to reduce the weight of steelwork. It incorporated advanced cable-spinning technology to lay the pair of 3-foot-diameter cables that make up the suspension system on each side of the bridge deck. The cables were spun in 209 working days with a labor force of over 300 men. In just seven minutes six loops were passed from end anchorage to end anchorage—a distance of a mile—and in an hour 100 miles of wire were spun. It required 217 loops to form a strand and there were 61 strands to make up each 3-foot-diameter cable. The accurate spinning of 107,000 miles of wire, which could encircle the equator four times, was an unprecedented feat of construction.

The tower steelwork was floated to the piers in 50-foot-long sections, and it needed 12 sections to complete each 635-foot tower. The sections of steelwork, after riveting and fabrication, were erected by huge derricks very rapidly. Workmen handled a million white-hot rivets, which had to be flung through the air hundreds of feet above the river, caught in buckets, and then driven into the steel with a pneumatic hammer, to complete all the joints of the tower sections. When the 180-ton saddles for the cables were in position on the top of each tower, it is estimated that 20,000 tons of steel had been used to build both towers.

The large caisson sunk for the west tower foundation was the largest cofferdam ever constructed. The east tower foundation was built on the shore and the massive cable anchorages were formed in the basalt rock on each bank, pouring concrete into a vast hole 220 feet by 209 feet by 130 feet deep blasted out of the rock. On such a natural site for a bridge, there was little that could go wrong.

The bridge was built in four years and opened on October 25, 1931, and is believed to be the only bridge ever built that was finished a year ahead of schedule! The bridge engineer was the great O.H. Amman and the architect was Cass Gilbert, the designer of the gothic Woolworth Building. The steelwork to the towers had been designed to be covered by concrete and granite panels, but, as the skeletal profile of the tower soared skyward, the natural beauty of the fabricated section fascinated and excited onlookers so much, that the bridge authority decided not to cover it, saving quite a tidy sum of money in the process.

LEFT ABOVE: A dramatic view of the distinctive steel lattice work of the George Washington Bridge tower.

LEFT BELOW: The bridge seen through the tower on the Manhattan side.

ABOVE: George Washington Bridge, New York, USA (1931).

Firth of Forth, Scotland (1890)

This was the most massive, monumental cantilever-truss structure ever built. Was it a masterpiece of engineering or a metal dinosaur that should never have been built? Some say it was so dated in its engineering that it cannot be hailed as a masterpiece of bridge design. But, on seeing it, one cannot fail to be moved by the sheer scale and muscular proportions of the trusses, which create such an awesome presence across a cold, weather-beaten landscape, that it evokes adulation and inspires admiration by everyone.

When it was finished, the Firth of Forth was the longest-spanning bridge in the world, beating the Brooklyn Bridge span by 100 feet. It established the cantilever-truss bridge as a serious competitor to the suspension bridge. There are still many fine cantilever-truss bridges standing today, but none were to match the spans of the Forth. The tragic collapse of the first Quebec Bridge expunged the growing trust in the efficacy of long-span cantilever construction. The honeymoon period for the long-span cantilever bridge was therefore short-lived.

The Firth of Forth Bridge would never have been built as a cantilever-truss construction if the Tay Bridge had not collapsed. Thomas Bouch's proposal for a long-span wire-cable suspension bridge had secured authorization from Parliament and had received funding, and work had actually commenced when, on December 20, 1879, the Tay Bridge, which he had designed, collapsed. Bouch was found negligent and of poor engineering judgment in the design of the main portals, and a year later he was asked to resign as designer of the Forth Bridge.

BELOW: Completing one of the balanced cantilever span sections.

inventor of prestressed concrete, was not satisfied with just making this discovery in his lifetime: he went on to exploit its unique properties, designing elegant shell roof structures, long-span exhibition halls, aircraft hangars, and bridges.

It was 1925, and cement technology was in its infancy: ready-mixed concrete, super-strength concrete, and pumped concrete had not yet been developed. Everything was mixed on site by crude drum mixers fed by workmen shoveling the materials into them—the large aggregates, the sand, and the cement. There was simple apparatus for weighing out the materials—the cement was supplied in bags, which had to be split open. On a windy day that fine grey power would get blown everywhere.

And yet, when Freyssinet submitted his winning design for the competition to build a bridge at Plougastel, he had the audacity to believe that he could produce a high-strength, rapid-hardening concrete with such crude technology. Using about 755 lbs/cu yard (450 kg/m³) of cement, he was able to achieve it! You need careful control of the mix proportions, computer-controlled weigh-batching, and specially formulated admixtures to make high-strength concrete from a ready-mix plant today. Freyssinet's proposal was akin to giving a surgeon a bread knife to perform a delicate operation on an eyelid—but he did it!

He needed high-strength concrete to construct the three identical concrete arches, each spanning 600 feet over the Elorn. The bridge was the longest span in France, beating the Garabit Viaduct, and for a short while it was also the world's longest concrete arch. The three identical arch spans have a cross section comprising a three-celled box, which is 31 feet wide and 16 feet deep at the crown. Spandrel walls built up from the arch support a double-deck truss which carries the roadway and a single rail track. The upper deck of the truss forms a 20-foot-wide roadway bordered by a 3-foot sidewalk on each side. The arch was concreted and reinforced in three stages. First, the slab for the lower chord was cast, then the box section walls, followed by the slab for the upper chord. The high early strength of the concrete was used to enable the prestressing to be applied after three days, before the concrete could shrink and crack. The other brilliant achievement in engineering the Plougastel Bridge is the story of how it was constructed. This is where Freyssinet saved money and time, and was able to beat the price of his competitors' designs. The bridge itself needed about 32,300 cubic yards of concrete along with forming materials, centering, and reinforcement. The difference in high and low tides can be as much as 26 feet in the Elorn, and, with wind speeds as high as 100 feet per second, it makes construction difficult in the river.

Instead of building the falsework for the centering across the river for all three spans in the traditional way, Freyssinet made a pontoon and floated out the centering for one arch only. This was removed and reused for constructing the other two arches in turn. The centering for the arch required only 260 cubic yards

1

3

Plougastel, Elorn, France (1929) This bridge represents the genius of one man's inventiveness, the exploitation of the plastic properties of concrete, the daring use of prestressing in concrete and an ingenious solution to building a multiple-arch span across a wide river.

Eugene Freyssinet's bridge over the Elorn at Plougastel was the finest bridge he had ever built and one of the truly great achievements in the art of concrete-bridge engineering the world has ever seen. This was the dawn of the age of concrete so masterfully heralded and understood by the bridge designs of Robert Maillart, but it was not until Eugene Freyssinet arrived that concrete-bridge engineering was taken to the limits of its capability. Freyssinet, the father and

BELOW: Plougastel Bridge as seen before a second bridge was built nearby.

of timber and was built on the Plougastel side of the river. A timber segmental arch was constructed with each end seated in a hollow concrete pontoon. The ends of the centering were tied together by cables whose tension could be regulated at both ends by hydraulic presses. Exploiting the tide, the concrete pontoons and the centering were towed out into position for the first arch on April 2, 1928. Once they were in position, the hydraulic presses lifted the centering to spring it to the correct level, and to the correct alignment with the help of the hangers at both ends of the centering. By August 7, when the concrete of the first arch had fully hardened, the centering was eased free and floated into the second opening. On January 19, 1929, it was floated into the third opening.

2

4

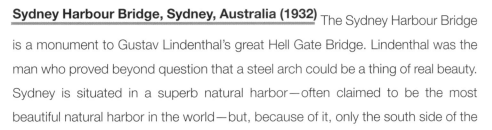

Sydney Harbour Bridge, Sydney, Australia (1932) The Sydney Harbour Bridge is a monument to Gustav Lindenthal's great Hell Gate Bridge. Lindenthal was the man who proved beyond question that a steel arch could be a thing of real beauty. Sydney is situated in a superb natural harbor—often claimed to be the most beautiful natural harbor in the world—but, because of it, only the south side of the city had grown, as the north side could be reached only by ferry or a circuitous ten-mile coastal road. Many proposals for a bridge had been made over the years, but nothing was workable since the ferries were providing an efficient service and the cost of a bridge was unjustifiable.

Then in 1922 John Bradfield, the chief engineer of the public-works department, published his proposal for a cantilever bridge across the harbor. But, before he finalized his design, he decided to go on a fact-finding mission to the United States and Europe, to study the world's long-span bridges. He was particularly impressed by the Hell Gate Bridge and upon his return to Sydney revised his design and copied the steel arch.

Sometime in 1923, after many bridge companies and engineers of the world had been invited to submit their design and tender proposals, the firm of Dorman Long were awarded the contract to build the Sydney Harbour Bridge. Not surprisingly, it was based on the steel arch that Bradfield had proposed, rather than a suspension bridge. There was a view that, with rock so near the surface, a suspension bridge would have been the better choice, but the competitive tender of Dorman Long suggested other-wise. Aesthetic grounds were cited for the selection of the steel arch, but it is also likely that a deciding factor was the ambitions of the Australians to boast the longest-arch span in the world.

The Sydney Harbour Bridge, like the Hell Gate arch, carries four railroad tracks, but in addition it carries a wide roadway and sidewalks. The dead weight of the span is 57,500 pounds per linear foot, while the Hell Gate was

ABOVE: Steel truss of the Sydney Harbour Bridge. 50,300 tons of steel were used in the making of the bridge.

52,000 pounds per linear foot. However, Sydney Harbour, built in 1932, spans 1,650 feet between abutments, while the Hell Gate, built in 1916, spans only 977 feet in comparison.

To complete the bridge 50,300 tons of steel was used, 37,000 tons of which was contained in the main arch. Special steel-fabrication yards were built close to the site of the bridge on the north bank. Panels of trusses were accurately assembled in 60-foot lengths. After riveting and painting, they were loaded onto barges and transported to a position directly under the bridge and lifted into place from the barge, by a 120-ton overhead traveling crane.

The erection of the main arch span was one of the greatest engineering feats of its time. The two halves of the arch were built out from each abutment, with each half held in position during erection by 128 steel cables, each 2.75 inches thick. These cables were attached to the end post of a truss, passed through a U-shaped tunnel cored into the solid rock foundation on the bank and then attached to the end post of another truss. Two traveling cranes were mounted on the top chord of the truss to carry the material for erecting the steelwork. Each crane was electrically operated and had a lifting capacity of 120 tons and weighed 565 tons. Between them, they erected all the steel for the steel arch.

To close the span the half-arches were lowered at the center, by gently slackening the anchorage cable at the links at the top of the end posts. Each cable was let out at both ends simultaneously. A team of six men on each half, in constant telephone contact, carried out the task, working two 12-hour shifts. It took five days to reduce the center gap to just 8 inches. The lower chord of each half was then connected by an 8-inch pin enclosed by a forged saddle fitted to the chord. The remaining tension in the cable was released very gradually, while the steelwork over the crown of the top chord was then assembled, joining both halves.

Now all that remained was to place the hangers from the main arch, the cross girders, and the bridge deck below. This work commenced from the center of the span with the cranes traveling backward toward the abutments. In May, 1933, the last piece of steel was placed for the bridge deck and the cranes were dismantled. The granite-clad towers that mark the ends of the arch and the intersection with the approach viaducts are an ornamental feature to convey the impression of mass. They do stabilize the arch but only at the springing line at the base of the tower.

Early in 1932 Sydney Harbour Bridge was opened and all the citizens of the city celebrated the completion of this monumental achievement, but they could not rejoice in the claim that this was the longest arch in the world. Four months earlier the Bayonne Bridge in New York was opened. It was 2 feet longer!

ABOVE: Sydney Harbour Bridge, Australia (1932).

RIGHT AND BELOW: Two views of Sydney Harbour Bridge, through the lens of photographer Grant Smith.

The Golden Gate Bridge, San Francisco, USA (1937)

The Golden Gate Bridge is one of the acknowledged wonders of the modern world, and the universal symbol of the modern suspension bridge. It stands for achievement, progress, and breathtaking imagery—it is the logo of a city that is known throughout the world, even though many people have never been to San

ABOVE: The "Span of Gold" across San Francisco Bay.

Francisco or know precisely where the bridge is situated. What is so great and so magical about this bridge? Is it the tall red towers, the mile-long cables that span the bay, or the setting of the bay with the city in the background?

The span of the Golden Gate is almost unbelievable at 4,200 feet between the towers. The distinctive stepped-back towers soar 746 feet in the air and are the tallest cable masts in the world. On pure engineering excellence the Golden Gate cannot be regarded as a very innovative or pioneering structure in its day—it just happens to be big. The sag of cables is excessive, the truss deck is very plain, and the architectural modeling of the towers by Irving Morrow is very affected, although it works extremely well.

Many engineers and bridge historians consider the 2,100-foot span of West Bay Bridge situated farther along the estuary to be a finer structure. But the Golden Gate has captured the hearts of the public and has become an institution, a monument to the enterprise and dynamism of San Francisco and the USA. It would be unthinkable to anyone today for such a bridge not to have been built in such a magnificent setting. And yet it did take a long time for the money to be found and the economics to be justified, before this bridge was built.

The need for a crossing between San Francisco and Saulite was not so pressing as the Transbay link between San Francisco and Oakland, which ferried millions of people every year across eight miles of open sea. The Transbay bridge, which included the West Bay and East Bay bridges, was built at a cost of $79 million between 1933 and 1936 and is still the longest high-level bridge in the USA at 43,500 feet.

When a scheme submitted by the consulting engineer J. Strauss to the city engineer and estimated to cost $27 million, it was received with some surprise. The bridge that Strauss had proposed was of a cantilever-truss design, supported

by suspension cables—it was also considerably cheaper than the budget of £100 million for a suspension bridge. Strauss argued that, for such a long span, a cantilever-truss construction like the Forth Bridge was too heavy, and a suspension bridge was not going to be rigid enough in such a windblown corridor, which also needed deep foundations in the sea. But some doubted that the design was worthy of a bridge that was to become the longest span in the world. Nevertheless, the city authorities were enthusiastic and canvassed support from the nearby counties and major industries to secure the necessary legislation to set up the Golden Gate Bridge Authority. An "opposition committee" led by prominent businessmen, taxpayers, and leading engineers was also formed to make counterclaims that the bridge was not needed and was impractical to build. After many years of political wrangling and arguing, it was left to the people of San Francisco to decide the fate of the bridge. On November 4, 1930, the public voted for the bridge to go ahead and construction finally began on January 5, 1933. In the intervening years Strauss's hybrid design was revised to a suspension bridge after consulting with Amman and other experts. It is the design that we see today.

ABOVE: The deep truss deck of the Tagus Bridge, seen here, was copied from the Golden Gate Bridge.

The most difficult part of constructing the Golden Gate Bridge was the pier foundations. This seems to ring true for all bridge types, from the smallest spans to the very largest bridges in the world. The pier on the west side encountered no difficulty, but the one on the other side, being in open sea, was unprotected from the elements and was vulnerable to the hazards of oceangoing ships. An access trestle bridge was built out to the pier 1,000 feet from the San Francisco shore. This is how men and materials for constructing the foundation and the superstructure of the pier accessed the site.

Not long after the trestle was built a ship was thrown off course in a fog and crashed into it, doing a lot of damage. And not long after that, in gale-force winds, 800 feet of it was carried away by the sea. It was rebuilt, and this time it was anchored to the sea bed and stayed in position for the duration of the construction, without further incident. When the caisson for the east pier had been towed into position inside a large concrete fender, it started to bob around like a cork on water under a heavy sea swell that developed through the night. The fender was there to

stop ships colliding with the pier when the bridge was built, and to protect the caisson under construction, but ironically the caisson was now threatening to batter the fender wall to pieces. The caisson was successfully towed out of the fender and the fender ring was enclosed and used as a cofferdam.

There were few serious incidents during the construction and it seemed that the Golden Gate Bridge was going to be blessed with good fortune. There is a saying among bridge workers that "the bridge demands a life"; loosely interpreted, that works out at one death per million dollars spent on the structure! From the start of the work only one life had been lost—that was until February 17, 1937, near the end of the contract. One of the scaffolds erected by the bridge paving contractor gave way, carrying with it 12 men and 2,000 feet of safety net, which was put there to stop men falling into the sea.

On May 27, 1937, a week of celebrations inaugurated the opening of the bridge. During this week the opening ceremony took place, at which the designer J. Strauss presented the bridge to the city. The nightly event of illuminating the bridge by floodlight was a great spectacle and the vermilion paintwork gave the structure the nickname of the "Span of Gold."

ABOVE: Sailing through the hangers of the suspension cable!

OPPOSITE: The distinctive stepped back towers of the Golden Gate Bridge rise 746 ft.

Severn Bridge, England (1966)

This is regarded as the world's first modern suspension bridge where the heavy truss deck—universally adopted after the Tacoma Narrows collapse—was abandoned in favor of a sleek aerodynamic box-girder design.

The bridge design was started in 1961 by a joint team of bridge engineers, Freeman Fox and Partners and Mott Hay and Anderson, with architectural consultancy provided by the Sir Percy Thomas partnership. The teams were appointed to design two of the longest suspension bridges in Britain and Europe at the time: the Forth Road Bridge, located just upstream of the famous rail bridge in Scotland, and the Severn Crossing near Bristol in the southwest of England.

Suspension-bridge design the world over had taken fright at the collapse of "Galloping Gertie," the nickname given to the first Tacoma Narrows bridge in Washington State, which began to twist, sway, and finally collapse under a moderate wind of 40 m.p.h. For the next quarter of a century all major suspension bridges were massively constructed with stiffening trusses, with older bridges like the Manhattan and the Golden Gate being retrofitted with deeper stiffening trusses.

The Forth Road (suspension) Bridge was designed conservatively with stiffening trusses, but the Severn, with a span of 3,240 feet, was formed with an innovative and radical streamlined deck section that minimized wind resistance and allowed huge economies of scale and material cost.

As early as 1953 the German bridge engineer Fritz Leonhardt had applied for a patent for a flexible stiffening truss which was suspended from only one supporting cable, and an A-frame cable tower, with rows of hangers arranged in a zigzag fashion, so that every four met at a point. This hanger arrangement helped to stabilize the suspension cable and bridge deck from wind oscillations. His later unsuccessful but radical competition design for the 1960 Tagus Bridge in Lisbon created quite an interest in the bridge world. Encouraged by the saving in cost that such a scheme potentially offered, Gilbert Roberts, the senior partner of Freeman Fox, sought Leonhardt's advice on streamlining the bridge deck of the Severn crossing and the feasibility of incorporating zigzag hangers.

BELOW: Construction of the aerodynamic bridge deck of the Severn Bridge.

LEFT: The stiff truss section of the Forth Road Bridge, which was built just before the Severn Bridge.

Wind-tunnel tests carried out by Freeman Fox proved conclusively that the combination of a streamlined deck with two supporting cables and diagonal hangers was just as effective for wind damping as a stiffening truss deck. It was the best way of dealing with the resonating, wind-induced oscillations, which had caused problems on the Forth Road Bridge, when the 512-foot-high towers started to sway.

The Severn's swift tidal flow created huge problems in building the foundations for the bridge pier. On the west pier workmen were able to work on the base only during two 20-minute periods a day, at low tide. After the towers were built and the cables spun, the 60-foot-long prefabricated "aerodynamic" metal deck sections were floated down river from the workshops and lifted by the hangers into position.

As well as a radical redesign of the deck, the towers of the Severn Bridge were designed as hollow box sections with internal stiffeners in the leg section, braced by deep portals, one immediately beneath the deck and two above it. This was a much lighter construction than designs based on the American cellular-plate

BELOW: A section of the radical aerodynamic bridge deck.

construction, and, although the towers (445 feet) were 70 feet shorter than the those on the Forth Road Bridge (512 feet), they used only half the amount of steel.

The Severn Bridge was opened to traffic in September 1966, five years after work had started. The Severn's revolutionary deck structure and tower fabrication have been adopted on many subsequent suspension bridges, notably the Humber Bridge in the UK, which became the world's longest suspension bridge in 1981 with a span of 4,624 feet.

BELOW: Recent pictures of the Severn Bridge, taken after the bridge had been repaired and repainted.

Sunshine Skyway, Florida, USA (1986) Disaster strikes on May 9, 1980

A thunderstorm had formed to the west of the Gulf of Mexico and was moving toward Tampa Bay. It was early morning on May 9, 1980, as Captain John Lerro, a member of the elite Tampa Bay pilots' association, was sent out on a 55-foot launch to bring in an empty cargo freighter, the *Summit Venture*, before it entered the tricky channels in the bay and passed under the Sunshine Skyway. It was just another routine job for Captain Lerro. He had piloted many big and small freighter ships successfully through the shallows and the sharp dogleg channels of the bay.

A mist, then a drizzle, followed by hard driving rain, swept very suddenly across the ship and the calm bay without warning. On the deck of the *Summit Venture* three lookouts kept watch for the all-important marker buoy that would tell Captain Lerro where to turn in the channel. Visibility was lost in the blackening sky and the colorless sea as rain lashed the deck. The ship was still moving steadily ahead, although by now the crew and pilot could see nothing but a screen of rain.

Suddenly Captain Lerro saw the Sunshine Skyway bridge looming ahead. He shouted the orders: "Hard to port—let go the anchor—ram the engines full Eastern!" But it was too late. The *Summit Venture* smashed into the tall support of the main span. As the ship hit the bridge, concrete and steel came falling down, some of it landing on the bows. Not only did the bridge deck collapse: six cars, a bus, and a truck fatefully traveling over the main span plummeted 150 feet into the sea. Captain Lerro called the Coast Guard repeatedly for help: "Mayday, mayday, mayday, Coast Guard—bridge down." The Florida Coast Guard and the Department of Transport were on the scene within the hour to help pick up survivors and to stop all traffic on the bridge.

In all 35 people plunged to their deaths that day, and only one person survived the tragedy. Wes MacIntyre, a vehicle maintenance man, was traveling over the bridge in the driving rain at 7.34 a.m. when suddenly his blue Ford Courier truck started to bob up and down and roll from side to side. He thought it was the wind and ignored it, but a few moments later he looked ahead and saw a ship in the water below and no bridge at all. His truck was airborne by the time he applied the brakes, and was dropping into the water. It bounced off the bow of the *Summit Venture* before sinking 30 feet to the seabed.

Wes MacIntyre returned to consciousness a moment later to find himself in his driving seat with the windows closed and water streaming into the cab of the truck. He took a deep breath, then forced the cab door open and swam to the surface. A crewman on the *Summit Venture* spotted him in the water and pulled him to safety.

In January, 1981, after much public debate and detailed investigation as to whether to repair or rebuild the original span, or construct an entirely new bridge, Governor Bob Graham announced the decision that an entirely new Sunshine Skyway would be built. It was to become one of the safest and most celebrated of modern bridges, and the longest cable-stay bridge in the USA.

The second Sunshine Skyway A bridge connecting St Petersburg and Clearwater across Tampa Bay has existed since 1954. The first bridge was built with an 846-foot, steel-truss-girder main span over the navigation channel. In 1971 an identical sister bridge was built alongside the old one, to increase the traffic flow. The original bridge was called the Sunshine Skyway in 1954 but, after the second bridge was added, it was officially renamed the Bill Dean Bridge after the chief of bridge design in Florida state. However, the public and press never really liked that name, preferring the magic of the original Sunshine Skyway.

ABOVE AND LEFT: A motorist view of the curved main span and the central cable stays.

Following the tragedy of the *Summit Venture*, the original bridges have now been replaced by an entirely new bridge running on a separate alignment, across four miles of Tampa Bay. The high-level approaches and the cable-stay navigation span were designed by the consultants Figg and Muller under their chief engineer, Jean Muller. The low-level approach spans were designed by the Florida Department of Transportation, while Parsons Brinckerhoff designed the bridge protection system. The dolphin bumpers and concrete islands placed around the main pier supports can take the impact of an 87,000-ton tanker traveling at 10 knots and not budge.

The cable-stay span is 1,200 feet in length, giving the roadway a maximum clearance above the water line of 193 feet. It contains 2.36 million feet of cable strand, which needed 2,500 gallons of bright-yellow paint to coat all the cable covers. The 70-foot-diameter pier support extends 175 feet from the water line to meet the bridge deck, and each one contains 13,000 cubic yards of concrete. The concrete pylon above the road deck tapers to a 50-foot diameter and finishes 431 feet above the water line. Each pylon supports two sets of 21 cable stays, picking up the bridge deck on each side of the span.

Construction work started in June, 1982, and the bridge was opened to traffic in April 1987 at an estimated cost of $245 million. In total the bridge contains enough steel to build a fleet of 746 Greyhound buses and enough concrete to form a 4-foot pathway from Pensacola to Key West. The 40-foot-wide, four-lane carriageway can accommodate 20,000 cars per day.

Because the climate is subtropical, hurricanes can occur in the Tampa Bay area. So the bridge has been designed to withstand a wind speed of 240 m.p.h. and a gust of 290 m.p.h. The highest wind speed ever recorded so far in the Gulf of Mexico, into which Tampa Bay flows, was produced by Hurricane Camille, with a wind speed of 190 mph in 1969.

Like many bridges in the world, the Sunshine Skyway holds a superstition. Some fishermen who fish under the approach spans of the bridge believe that the body of a construction worker lies buried in the concrete in one of them—a victim of a construction accident. Some late-night motorists claim to have seen a woman on the main span waving frantically for help, but just as they slow down to stop the apparition is gone.

Pont de Normandie, Honfleur, France (1995)

Another giant leap for progress in bridge engineering was taken when the Pont de Normandie was opened to traffic in 1994. We need to go back to 1931 when the George Washington Bridge smashed the longest-span record by over 60 percent, to find a jump to compare to the cable-stay span of the Pont du Normandie. The Pont de Normandie, which has a main span of 2,800 feet, has beaten the record held by the Skarnsundet Bridge of 1,740 feet set in 1991, and the Quingzhou Minjang Bridge built in China in 1996, with a span of 1,980 feet, by a big margin.

The geology of the site and the river bed and the light traffic flows the bridge was to carry were the two critical factors that tipped the balance in favor of a cable-stay bridge rather than a suspension bridge. The third factor, hardly ever admitted on record, was the emotive one of nationalistic pride in wanting to build the world's longest cable-stay bridge! The solution of a suspension bridge was ruled out because enormous anchorages would have to be built into the soft alluvium, since the site was devoid of good natural ground support.

From end to end the bridge is a mile and a quarter long, with extensive approach viaducts on both ends of the central span. The main span section under each tower, and the side span connecting the viaduct on both sides of the bridge, were designed as prestressed-concrete box sections, and built using the balanced cantilever method. The side spans and the main span section under the towers provide the rigidity for controlling the deformation of the large central span. Steel box girder rather than concrete was preferred for the 2,050-foot central span section, to reduce the dead weight of the span.

On July 5, 1994, the last of the steel box-girder sections weighing 200 tons was slowly lifted above the water of the Seine. The 32nd segment, which measures 75 feet wide, 16 feet deep, and 65 feet long, had taken three hours to move the 165 feet from the water level to the deck. Now all that remained to be

ABOVE: Temporary cables are attached to help support the main deck structure.

BELOW: Elevation drawing of the Pont de Normandie.

FOLLOWING SPREAD: Pont de Normandie is now as much a part of Normandy heritage as Calvados and Camembert.

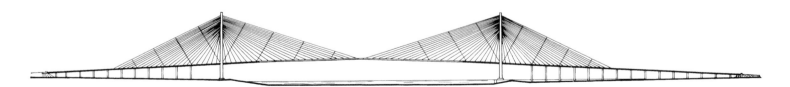

ABOVE LEFT: The cable stays stretch down from the pylon and are then fixed to the bridge deck.

ABOVE RIGHT: Looking up at one of the main pylons, and the web of permanent and temporary cable stays that sweep down to the deck.

done was to weld the box girder to the adjacent segment and then to join the two huge cantilever halves of the bridge, by inserting a small 8-foot-long keying section. The Seine can now be crossed near Le Havre without making a 30-mile detour to cross at the Tarcanville Bridge. It had taken four years of financial planning and technical study and six years of construction to complete the bridge. The two A-frame concrete cable towers standing guard over the estuary are as tall as the Montparnasse Building in Paris, France's tallest building, while the main span—which is held up by 184 gigantic stays—is as long as the Champs Elysées.

The stay cables were installed at the rate of two every day, with the geometry of the entire structure having to be adjusted constantly, as they were fitted. During the progress of construction computer specialists had to integrate and process up to 36 topographical measurements taken every day and assimilate more than 5,000 plans and sets of calculations to check the movement and stress effects on specific parts of the structure. Without the fantastic progress that has taken place with information technology and computer science in the past decade, this bridge would never have seen the light of day.

In windy squalls the construction engineers found that the longest cables began to vibrate at the same rhythm as the bridge deck, setting up dangerous harmonics. There was a risk that the whole bridge would begin to resonate like a gigantic harp. To prevent this happening, 32 bundles of transverse wire ropes were installed across the cable stays to dampen these oscillations. Stay cables can slacken under their own weight, just like a washing line. When the wind blows they tend to vibrate. If they are tied to one another, the livelier ones are calmed by the others.

Before the bridge was officially opened on January 20, 1995, it was given some punishing tests. First a fleet of 80 35-ton lorries maneuvered on the deck so that the vertical sag, the horizontal movement of the towers, the distortion, and load on the stay cables could be measured. The results tallied almost exactly with the engineered calculations. Then, a few days before the opening, a powerful oceangoing tug was deliberately moored to the deck, and then pulled on the mooring ropes before suddenly releasing its 140-ton strain, but the bridge deck did not waver.

"To send more than 70,000 cubic meters of concrete and 19,000 tons of steel arching into the sky between Honfleur and Le Havre would have been unimaginable twenty years ago," says Michel Virlogeux, lead designer of the Pont de Normandie. "This project has been a real scientific adventure from start to finish." It may take till the year 2001 and beyond to recover the $456 million that has been invested in the construction of this great bridge, not that the people of Normandy will mind, because they love the bridge. It has already become part of their heritage, like Camembert and Calvados.

ABOVE (CLOCKWISE): The anchor plate that connects the cable stays to the pylon head; Attaching the transverse cable ties to dampen any cable vibration in high winds; Abseilers checking the cable cover shields.

"This project has been a real scientific adventure from start to finish," says Michel Virlogeux, chief design engineer of the Pont de Normandie.

5

A chapter of disasters

No history of bridges could be complete without knowing something about the many bridge collapses, loss of life, and the catastrophic failures that have darkened the horizons of bridge-building progress over the centuries. The most dangerous period for a bridge is when it is being built: temporary supports can fail, while human error can lead to neglect of a critical operating procedure or an underestimate of the stresses on the bridge. Occasionally forces of nature can be just too overpowering and overwhelming for a structure to resist. And just occasionally a bridge can be damaged by accidental impact from a ship or the derailment of a moving train. These are unforeseen events that the bridge designer would not have allowed for.

Many people may be under the illusion that bridge failures do not happen nowadays, but nothing could be farther from the truth. Nineteen ninety-eight was one of the biggest boom years for bridge building in the world, with some of the longest, tallest, heaviest bridges the world has ever seen being built.

It was also a bad year for bridge accidents. During 1998 there was a report almost every month of an accident or collapse of a bridge somewhere in the world—and there were fatalities.

ABOVE: Milford Haven, Wales—steel box girder span buckles during construction in June 1970.

OPPOSITE: Collapse of Cypress Viaduct, after the San Francisco earthquake of 1989.

LAUNCH SYSTEM SCRAPPED
after Japanese bridge fall

KURUSHIMA BRIDGE, JAPAN.

Japanese bridge engineers are conducting urgent safety checks on their most widely used incremental deck launching technique following the release of a scathing report on the country's worst bridge accident in decades, which killed seven workers. The accident occurred in June on one of the world's largest bridge projects, the £5.5 billion [approximately $8.8 billion] Kurushima crossing, near Imabari, southern Japan. Sections of temporary steel platform supporting an already launched side span were being dismantled and lowered on cables. Three cables gave way, tipping the 50 ton platform section and plunging workers on it 60m [196 feet] to the ground.

(from *New Civil Engineer*, September, 1998)

Injaka bridge collapse—

South African experts probe incremental bridge launch

Fourteen people died and thirteen were seriously injured when two of the Injaka Bridge spans collapsed during construction. The accident happened when a 27m [88 feet] long steel nose girder attached to the front of the first deck segment had just been pushed on to a temporary bearing on the second pier when suddenly both spans collapsed, dropping workers and a party of visitors 30m [98 feet] to the ground. A full scale investigation into tourteen people died and thirteen were seriously injured when two of the Injaka Bridge disaster has been launched by the South African Department of Labor, backed by the Police and Health and Safety Inspectorate.

The bridge collapse was all the more tragic as a party of guests invited to the site by the consultants VKE Engineers who designed the bridge, were injured, some fatally. What was intended to be a special occasion to celebrate the completion of the second span turned in moments into a scene of destruction. Among the dead was the bridge's designer 27 year old Marlieze Gouws, described by colleagues as a competent young engineer who had gained an honors degree with distinction.

(from *New Civil Engineer*, July, 1998)

BRIDGE COLLAPSE

(from *Civil Engineering International*, January, 1998)

One man was killed and two men injured when a temporary bridge collapsed during demolition near the town of Iaeger, West Virginia. Director of West Virginia highways construction Bob Tinney said the 50m [165 feet] long central span fell as contractor Battleridge Companies was preparing to lift it out of position. "The contractor had a crane tied to the span and I understand that connecting bolts were being taken out at the time of the accident," said Tinney. He added that the company had worked regularly for the state and had a good safety record.

Let's now go back in time to record the events of some of the worst bridge disasters in the history of bridge building.

Wheeling Bridge, Ohio, 1854 The collapse in 1854 of the 1,000-foot-span Wheeling Suspension Bridge, the longest bridge in the world, was a terrible lesson that was lost to the profession, because history was to repeat itself almost 100 years later. David Steinman recounts in his book *Bridges and Their Builders* a remarkable eyewitness account of the Wheeling collapse, which was reported in the Wheeling *Intelligencer*:

With feelings of unutterable sorrow, we announce that the noble and world renowned structure, the Wheeling Suspension Bridge, has been swept from its stronghold by a terrific storm … At about 3 o'clock yesterday we walked toward the Suspension bridge and went upon it, as we frequently have done, enjoying the cool breeze and the undulating motion of the bridge … We had been off the [bridge] only two minutes and were on Main Street when we saw persons running toward the river bank: we followed just in time to see the whole structure heaving and dashing with tremendous force.

For a few moments we watched it with breathless anxiety, lunging like a ship in a storm; at one time it rose to nearly the height of the tower, then fell, twisted and writhed, and was dashed almost bottom upward. At last there seemed to be a determined twist along the entire span, with about one half of the [bridge deck] being reversed, and down went the immense structure from its dizzy height to the stream below, with an appalling crash and roar. Charles Ellet completed his greatest work, the Wheeling Suspension Bridge, over the Ohio River in 1848. Six years later, on May 17, 1854, the great bridge was destroyed by the wind. The bridge that Ellet had designed was quite capable of supporting its own weight and the loading from bridge traffic of horses, carts, wagons, people, and cattle. It was robust enough to resist the force of a considerable wind, but not the vibration patterns set up by the wind on the deck, which in turn could cause the cables to sway. The Wheeling Bridge illustrated how little the principles of aerodynamic stability were understood during this period, with the exception of John Roebling. He had worked out why the Wheeling Bridge had failed and seemed to have anticipated such problems in his designs, introducing traverse or cable stays between the bridge deck and the suspension cables.

No one was killed or injured. The bridge was later rebuilt by Charles Ellet with transverse stays and a stiffer bridge deck structure, and reopened in 1860.

ABOVE: Wheeling Bridge, Ohio, USA, after it was rebuilt.

Ashtabula, Ohio, USA 1876 In America in the nineteenth century no railroad promoter could afford to build bridges entirely in cast iron, however strong and efficient it could be. A railroad could reap handsome profits by building a rail track across hundreds of miles of the countryside just adequate for carrying a steam locomotive pulling a few light rail cars. A rail bridge had to be built at the smallest cost, capable of supporting the loads of a slow-moving train and nothing more. An iron arch, while being cheaper than a tubular-iron truss, was still more expensive than a composite wood and iron truss. So the wood and iron truss was what the railroad settled for. Many of their bridges were based on the Howe truss—it was one of the more popular designs. But, as the trains became heavier and the locomotive more powerful in the 1850s, many Howe-truss bridges began to collapse.

In 1865 the Lake Shore & Southern Michigan Railroad faced the problem of replacing an important bridge at Ashtabula, Ohio, over Ashtabula Creek. The creek was shallow but it ran through a gorge 700 feet wide and 75 deep. A new bridge was built, made entirely of wrought iron in the form of a modified Howe

ABOVE: Ashtabula Bridge in service before it collapsed.

truss. It represented a new approach to rail-bridge construction in the USA. It had a span of only 150 feet because the railroad company had built up the embankment on each side of the gorge.

Eleven years passed without incident until the night of December 29, 1876. The Pacific Express, a westbound 11-car train with double locomotives, had started to travel across the Ashtabula Bridge in a blinding snowstorm. Halfway across, Dan McGuire, the driver of the front locomotive, felt a huge shudder and the sensation of the train going uphill. He reacted by immediately opening the throttle to make the locomotive surge forward and, as it did, he heard a loud grinding sound. It was the tender of his locomotive scraping against the bridge abutment.

In the next moment there was a mighty crash as the second locomotive smashed into the same abutment. McGuire's locomotive then sped forward along the track and off the bridge as the coupling bar connecting the tender to the second locomotive was sheared. He brought the locomotive to a halt and then backed it along the track. By the time he jumped off and had run to the edge of the embankment, a ghastly illumination was glowing through the snowstorm, down in the gorge. The rest of the train, composed of two express cars, two baggage cars, two passenger coaches, a smoking car, a drawing room car, and three sleeper cars, had fallen into the gorge, crashing into one another. The wood-burning stoves in the cars had set light to their wooden hulks and were ablaze.

Of the 90 people killed, most of them died instantly from the crash as the cars plunged into the gorge. Those who were not crushed to death were uncertain whether to remain inside the cars and risk being burned, or face being frozen in the icy waters of the Ashtabula as they got out. Out of the coaches, hands and arms were thrust forward and slowly a trickle of confused and terrified passengers clambered out and staggered up the embankment free of the fires. It took a while for help to arrive.

This collapse of a wrought-iron bridge reverberated throughout the US. Newspapers and magazines carried stories about the collapse: the *Iron Age* magazine voiced the fears of the people, saying, "We know there are plenty of cheap, badly built bridges, which the engineers are watching with anxious fears, and which, to all appearances, only stand by the grace of God!" While the *Nation* summed up the apathy for better safety in design declaring, "By such disasters and by shipwrecks are lives in these days sacrificed by the score, and yet except through the clumsy machinery of the coroners' jury, hardly anywhere in America is there the slightest provision made for an inquiry into them."

The coroner put the blame of the Ashtabula disaster down to the designer of the modified Howe truss, Amasa Stone, and on Charles Collins, the chief engineer of the rail line, who survived the disaster by being in the last coach. Collins was unfairly blamed and was made a scapegoat by the press. He was a sensitive, gentle person, who suffered horribly with personal agonies of guilt and finally committed suicide. On the other hand Amasa Stone, a more ebullient, arrogant individual, defended the design of the bridge and his integrity very robustly.

The Ashtabula disaster was down to the lack of knowledge of cast-iron behavior—its low resistance to tensile forces, the need for additional diagonal

BELOW: A chart of the collapsed bottom chord of the truss, where it fell in Ashtabula Creek, as observed about three weeks after the disaster took place.

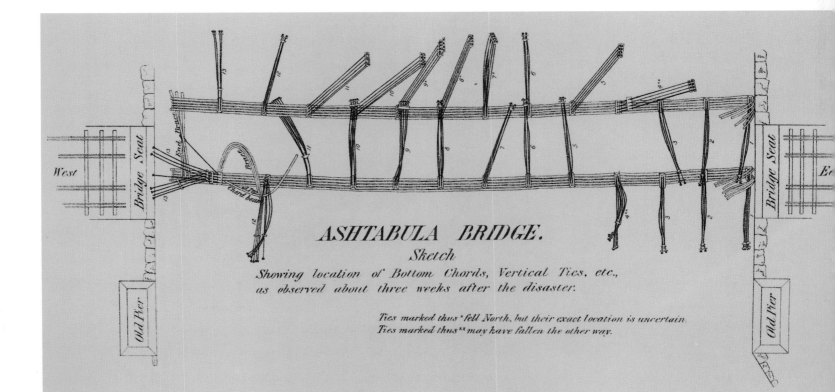

ASHTABULA BRIDGE.
Sketch
Showing location of Bottom Chords, Vertical Ties, etc.,
as observed about three weeks after the disaster.

Ties marked thus * fell North, but their exact location is uncertain.
Ties marked thus ** may have fallen the other way.

bracing of trusses, the brittle failure of wrought iron under repeating tensile loading—and the cavalier fashion by which the railroad cut corners and often disregarded the design capacity of a bridge. It was the derailment of the train, however, in the heavy snowfall on the track, that shifted most of the weight of the train to one side of the truss, causing the stresses to be reversed and to go into tension, which then caused the structure to buckle and collapse.

There were many truss bridges built in the US that were just like the Ashtabula. Two years later a wood-and-iron Howe truss at Tarifville, Connecticut, fell under an excursion train, killing 17 people. There was no derailment: the bridge simply collapsed. During the 1870s no fewer than 40 bridges a year fell, half of them purely timber structures. But the iron-bridge failures made the headlines, because they were the biggest and their collapse resulted in the greatest loss of lives. In the ten years after the Ashtabula tragedy, 200 bridges collapsed, many involving major loss of life.

It marked the end of the cast-iron bridge era in the US and the introduction of major reforms in bridge safety standards, of a state bridge inspectorate, and the requirement by law that a bridge design specification should be passed by the ASCE, the American Society of Civil Engineers.

The Tay Bridge disaster, 1878, Dundee, Scotland

Two years after Ashtabula, and almost exactly to the day, a worse disaster shocked the bridge world and dented the pride of a great bridge-building empire.

Thomas Bouch, the designer of the completed Tay Bridge, was busy designing the biggest bridge in the world, the Firth of Forth, when he received the awful news "that the bridge was doon." Thomas Bouch was until that moment the most renowned of living engineers in Britain. In June that year he had been honored when Queen Victoria and her royal train stopped for a while at Tay Bridge Station to receive an address of welcome and to meet with leading officials, including Bouch. The Queen's train had made a special detour from Balmoral on its way down to London to cross the newly opened Tay Bridge. Shortly afterwards Bouch received his knighthood from Queen Victoria, standing alongside Henry Bessemer, who was the inventor of a process for converting iron into steel, at Windsor Castle.

Although the Tay Bridge was an extraordinarily long bridge and the biggest bridge in overall length in the world, it had a fairly standard design, and did not feature any new ideas or construction techniques. Each big lattice girder in wrought iron was riveted together on the south shore, then floated out on heavy barges and raised onto the piers by hydraulic jacks. Several of the foundations were sunk by pneumatic caisson.

The completed bridge was a thin, long ribbon reaching out from the shore on slender supports rather like a fragile seaside pier, with the distinctive high girder spans in the middle to provide clearance for shipping. There was nothing spectacular or inspiring about the bridge and there were no adverse comments about its safety, with the exception of a chance remark made by Major General Hutchinson of the Royal Engineers, when inspecting and testing the bridge to issue its structural worthiness certificate on behalf of the Board of Trade. In his report Hutchinson made some minor criticisms of the bridge and recommended that trains be restricted to a maximum speed of 25 m.p.h. At the end of the report he made a casual reference to the wind: "When again visiting the spot, I should wish if possible to have an opportunity of observing the effects of high winds when a train of carriages is running over the bridge." This observation was never made in the final inspection as Hutchinson fell ill shortly afterward and his replacement officer did not attach any importance to it.

It was already dark when the storm of Sunday, December 28, 1878, struck Dundee. The strength of the storm increased over the next few hours into a gale as the wind speed climbed from Force 10 to Force 11 approaching 100 m.p.h. The gale blew across the open stretch of the Firth of the Tay, buffeting the bridge. It was just gone seven o'clock in the evening when the mail train from Burnt Island arrived at Tayport and slowed down to three miles an hour to be given the signal to cross the bridge. Thomas Barclay, the signalman, handed the fireman the clearance baton, and then hastened back to his hut to send a wire signal to the north shore to let them know the train was on its way.

From his hut Thomas and a colleague, John Watt, watched the train's lights as it made its way over the bridge in the driving rain and the howling gale. There was a spray of sparks from the wheels and then three flashes and one big flash. Suddenly Watt could not see the taillights of the train.

BELOW: Illustration of the Tay Bridge shortly after it was built.

"There's something wrong with the train," said Watt. Barclay was not so sure and thought the train was hidden in the curve of the bridge and was temporarily blocked from view. After a while, when no lights reappeared on the far side, Barclay immediately rang the bell to signal the north shore, but there was no answering bell. He tried his speaking tube but got no reply.

Thomas and Watt then ran out of the hut into the gale and started to scramble across the bridge. They had to get on their hands and knees for fear of being blown into the estuary, but after 20 yards they gave up and turned back. They climbed off the bridge down onto the bank at the water's edge in the rain-sodden blackness in the hope that they might see something. Incredibly, for a brief moment, the moon came out and to their horror they saw that the middle section of the bridge where the 13 high girder spans should have been had disappeared.

The two men turned and scrambled up the banks to raise the alarm that the "bridge was doon." A tense crowd had gathered in the harbor on the south shore, aware that a disaster had happened because they too had seen sparks, flashes, and columns of white spray leaping up from the black waters. A ferryboat was ordered out but returned at midnight with no sign of any survivors despite carefully negotiating the pier stumps of the bridge.

Farther downstream there were reports of mailbags being washed ashore. The news was telegraphed to London and the next day the whole world was in shock. Seventy-five people died in the disaster but only 46 bodies were ever recovered; 29 have remained buried in the Tay to this day.

BELOW: Looking for survivors in the icy waters of the Tay.

What caused the disaster? Two points of view emerged immediately. The religious extremists blamed the railroad for running trains on Sundays, saying it was the "hand of God" determined to guard the Sabbath. The business community and many others thought the man to blame was the designer.

An investigation was conducted by a court of experts appointed by the Board of Trade, and commenced on the spot in Dundee, as divers searched the Tay for bodies. They interrogated dozens of witnesses, listened to the facts, the explanations, and the theories of what may have caused the collapse from everyone including Sir Thomas Bouch. Several damning things emerged concerning the quality of the cast iron in the girders and the piers, which had been found with "Beaumont Eggs." This is the practice of filling holes found in the cast iron with a mixture of beeswax and iron filings. It turned out the inspector responsible for the maintenance of the bridge was inexperienced and incompetent.

Locomotives were often allowed to race across the bridge exceeding 25 m.p.h. to overtake the ferry. But the most appalling discovery was when they found out that Sir Thomas Bouch had scarcely considered the full effects of the wind in the Tay when designing the high girders of the bridge. Bouch had allowed for a wind pressure of 20 pounds per square foot acting on the bridge. His calculations were based on wind tables prepared by John Smeaton in 1759, about 120 years before.

ABOVE: What remained of the support legs and the high girders after the Tay Bridge collapsed on Dec 28, 1878.

Smeaton's tables were not inaccurate: they were simply not appropriate for an open estuary like the Tay, where wind gusts would greatly exceed the steady uniform pressure assumed by Smeaton. Bouch had also consulted with the Astronomer Royal, Sir George Airy, for advice and he confirmed that "the greatest wind pressure to which the plane surface like that of a bridge would be subject to is ten pounds per square foot."

In summing up Henry Rothery, Her Majesty's Wreck Commissioner and one of the members of the three-man court, stated in his report, "I do not understand [why] my colleagues differ from me in thinking that the chief blame for the casualty rests with Sir Thomas Bouch, but they consider it is not for us to say so …" And he went on to explain his reasons: "Engineers in France made an allowance of fifty-five pounds per square foot for wind pressure and in the United States an allowance of fifty pounds was made."

Sir George Airey had written to the court and advised them that in his opinion a wind pressure of 120 pounds per square foot should be assumed for bridges, not ten pounds per square foot! Rothery then spelled out the blame: "The conclusion ... is that this bridge was badly designed, badly constructed, and badly maintained and that its downfall was due to inherent defects in the structure which must sooner or later have brought it down. Sir Thomas Bouch is, in our opinion, mainly to blame."

The bridge had been strained by previous gales and by the trains that ran on it at excessive speeds. The wind acting on the bridge during a gale is not a single pressure, but a series of gusts. These gusts would have a greater overturning effect on the bridge than the measured pressures of a Force 11 gale. The twin trusses and flooring of each high girder span would present a large exposure area, with the force from such gusts enough to possibly collapse the bridge, particularly as the span may not have had sufficient lateral stability, after being shaken by previous gales and the vibrations from passing trains.

It is likely that, as the train traveled over the juddering high girder spans and was slammed broadside by the high winds, the riveted connections of a girder and pier came loose and something broke. A few moments later the pier collapsed, and a girder fell, then girder after girder and pier after pier just crumpled like a deck of cards, plunging the locomotive, the passenger cars, and the people 88 feet down into the freezing waters of the Tay.

Sir Thomas Bouch was destroyed: he was dismissed as the engineer of the Forth Rail Bridge and died from pneumonia not long afterwards, a bankrupt and broken-hearted man. But was it really all his fault?

The Quebec Bridge disaster, 1907, Canada The arrival of the twentieth century heralded a new dawn for bridge construction. No longer was bridge building dominated by the masonry arch and the timber and wrought-iron truss. The arrival of steel provided a real battleground for bridge forms—on one side there was the suspension-bridge lobby championed by the mighty Brooklyn Bridge, and on the other the massive cantilever truss of the Forth Rail Bridge. Was the cantilever-truss arch a more stable structure than a suspension bridge? Which bridge form was quicker to erect, cheaper on cost, safer under wind loading, and more able to support a fast-moving railroad?

The American railroad companies preferred the cantilever-truss bridge to the suspension bridge because they felt its rigidity made it ideal for the heavy loading of the railroads. When a new rail bridge was proposed over the St Lawrence River valley in Quebec, Theodore Cooper, one of the most eminent bridge engineers, was asked to design a long-span cantilever bridge. It was going be the longest-span bridge in the world, beating the span of the Forth Rail Bridge by 100 feet.

However, the first estimate of the weight of steel for the bridge pointed to an enormous cost, and Cooper and his team were put under pressure by the rail board to use all their resourcefulness to keep down the tonnage. The design that

OPPOSITE ABOVE: The critical cantilever span of the Quebec Bridge being erected.

OPPOSITE BELOW: Tangled heap of the cantilever span lies strewn on the shore of the River Lawrence.

Cooper evolved, although inspired by the Forth Bridge cantilever, differed in construction concept because it had only one main span and two end spans. The Forth was built with the cantilever spans in balance, and built out equally from the central supports. The relatively short end spans at each end of the Quebec Bridge had to secure the longer length of cantilever main span during construction.

Work was going along well, the end spans had been built, and the south cantilever arm was nearing mid-span, when the resident engineer, Norman McClure, sent Cooper a telegram telling him that the cantilever was starting to deflect downward by a fraction of an inch. McClure sent more telegrams to Cooper on subsequent days, as the deflection increased, urging him to visit the site. Then on August 27 he sent a final telegram to Cooper saying "erection will not proceed until we hear from you and from Phoenixville." The Phoenix Bridge Company, who were erecting the bridge, were based in Phoenixville and that's whom McClure was referring to in the telegram. They had responsibility for the bridge works during its construction and naturally had to be consulted.

The next day Cooper replied, ordering an immediate investigation into the deflection, aware that he may have pared the structure down too drastically to save on the cost of steel. His message did not state that all work should be halted, but neither did it say carry on regardless. McClure decided it was best to stop all construction work that day and left for New York to see Cooper (who was 70) in person. A contractors' superintendent on site, not aware of any potential danger—one assumes he was not subject to directives by the resident engineer—sent the men to work on the cantilever arm as usual the next day, August 29. Even though the deflection had now increased visibly, a crane was moved out onto the span during the day.

On the evening of August 29, 1907, just before work stopped for the day, the sound of tearing metal filled the air, signaling the worst bridge construction disaster ever recorded. The incomplete cantilever arm of the south span broke away and 19,000 tons of steel crashed into the St Lawrence river with 86 men on it. Eleven men luckily survived the tragedy.

Insufficient stress-analysis theory, construction knowledge, and engineering mathematics—these were the real deficiencies with the bridge, not the fact that the steel had been cut right down to the bone. Some of the compression elements in the truss of the cantilever arm were subject to a large squashing force that tended to bend them outward, making them buckle and ultimately give way. The rupture point of the Quebec Bridge was at the connection to the main chord. The joint actually buckled at the connecting plate, which was not made rigid or strong enough. It had only two connecting rivets when it should have had eight, according to the inquiry team. In fact the joint had only 30 percent of the strength of the compression member it was connecting.

It was clear that the accepted design rules at the time, based on the empirical design of compression members and connecting plates of much smaller structures, was unsuitable for very large-scale structures. Theodore Cooper, one of the finest bridge engineers of his time, who had worked as Eade's right-hand man on the St Louis Bridge, died shortly after the event, a broken man. More scientific study was necessary to understand the behavior of buckling in compression members and the proper detailing of construction joints.

That said, ten years later in 1916, by which time the second cantilever Quebec Bridge had been redesigned with K-bracing and was nearly built, a second tragedy happened. The central suspended span was being jacked up into position from a floating barge, when one of the jacks failed with the steel truss 15 feet in the air. Five thousand tons of steel crumpled and fell into the St Lawrence, killing 11 workers.

Quebec Bridge, with two horrendous collapses costing 87 lives, had taken over ten years to finish, just to claim the world's longest span by 100 feet. Was it worth the price?

BELOW: The rebuilt Quebec Bridge, 1917 with the additional "K" bracing.

The Tacoma Narrows Bridge failure, 1940, Pugent Sound, USA

With the Golden Gate and Oakland Bay bridges recently completed, America led the world in suspension-bridge construction, and now embarked on a crusade for even more graceful, slender, and daring span designs. None was more elegant than the span of the Tacoma Narrows Bridge over the beautiful Pugent Sound in Washington State, which opened to traffic on July 1, 1940. From the very day it opened there was something extraordinary about it. Motorists crossing the 2,800-foot span frequently saw the car ahead sink into the road and then rise again. No one was alarmed, since the engineers assured drivers that the structure was safe, even though the span undulated in breezy conditions. Drivers began to really enjoy the novelty of the gentle roller-coaster ride across the flexible span of the bridge. It became quite an attraction and soon acquired the nickname of "Galloping Gertie." Toll revenues from the bridge ran higher than expected—everyone was pleased.

Four months later, on November 7, with a moderately high wind blowing at about 44 m.p.h., the span suddenly went into an alarming series of rolls and pitches. One side of the road dipped and the other side rose, as a ripple wave traveled the length of the span. The bridge was closed to traffic. One motorist stopped his car partway across as the deck motions became quite terrifying. He got out of the car but his dog would not leave it. After a few moments, he walked back alone. A few minutes later, some suspension hangers tore loose and a 600-foot length of concrete deck fell into the sound. The remainder of the deck now whipped and thrashed about in a frenzy, until it too ripped free of the hangers, sending the whole of it into the Sound. With it went the car and the dog.

ABOVE: The final moments before "Galloping Gertie" plunges into the Pugent Sound.

RIGHT: A 600 ft section of the Tacoma Narrows bridge is torn away, as the structure writhes and twists uncontrollably.

LEFT: The second Tacoma Bridge was designed with a deep truss deck to prevent it from vibrating in high winds.

The Tacoma collapse was the most spectacular bridge failure of all time. Every moment of its dance of death was filmed and recorded by Professor Farquharson of the University of Washington. The film footage of the collapse became a newsreel classic. The designer of the Tacoma was blamed for its collapse. He was Leon Moissieff, who was responsible for the design of the beautiful Manhattan Suspension Bridge, and a leading consultant on the Golden Gate, Oakland Bay, and Bronx Whitestone bridges with Othmar Amman and many others. At 68, he was one of the most experienced suspension-bridge engineers in America. When he submitted his slim, plate-stiffened deck-girder span for the Tacoma Bridge, based on the success of the Cologne Bridge designed by Fritz Leonhardt, and which had been used by Amman on the Bronx Whitestone Bridge, nobody argued that it was too slender or unsafe in high winds. Quite the reverse was true—bridge engineers admired and raved about Moisseiff's streamlined design.

Steinman summed up the tragedy of the Tacoma poignantly: "The span failure is not to be blamed on Moissieff alone; the entire profession shares in the responsibility. It is simply that the profession had neglected to combine and apply in time, the knowledge of aerodynamics and dynamic vibrations with its rapidly advancing knowledge and development of structural design." The lessons of the Wheeling Bridge had been forgotten in the intellectual challenge of designing long-span structures during the 1930s and 1940s.

Thereafter, the suspension-bridge decks in the United States were designed with deep stiffening trusses to avoid dynamic instability in high winds. The Tacoma Bridge was rebuilt later with a deep truss deck, while the Golden Gate, the Bronx Whitestone, and many others were retrofitted with stiffening trusses.

West Gate, Australia and Milford Haven, Wales Failure of steel box girder bridges in the 70s

In the 1970s the world's attention was focused on the collapse of four steel box girder bridges during construction. The bridges were in Vienna over the Danube, in Koblenz over the Rhine, in Milford Haven, Wales, where four people were killed, and in Melbourne over the Yarra River, which killed thirty-five people.

It has become increasingly clear over the centuries as stronger materials are developed and new construction technologies evolve, that the limiting factors are not technological nor economic constraints, but the human ability to effectively communicate with one another and to decide on priorities. The more advanced the technology and the greater the economic pressure to save money, the more critical become the consequences of human actions or the lack of them. This was how bridge experts summed up the cause of failure of a series of steel box girders bridges in the 1970s, the worst of which was the West Gate Bridge collapse which occurred only four months after the Milford Haven collapse.

On October 15, 1970, the west cantilever section of the main span of the West Gate bridge suddenly crashed, bringing down 1,200 tons of steel on to workmens' huts below. Thirty-five people lost their lives, many more were seriously injured. How could such a collapse have happened only four months after a similar box girder span buckled and collapsed in Milford Haven, Wales? Both had been designed by the same team of consulting engineers.

One of the problems of the steel box girder is that during construction as the span cantilevers further out, the tension and compression stress in the beams is often greater than that from the eventual loads that they have to carry. As a result of this the deflection or sag at the tip of the cantilever may become very pronounced. Usually the ends of the cantilevers have to be hydraulically jacked up to align and connect the two cantilevers. This is a very time consuming operation. For the West Gate box girder, to make construction easier, each cantilever box section was prefabricated in two halves, and then hoisted into position and bolted together. A misalignment occurred between the two halves of the west cantilever section when it had reached 300 feet. It was decided to ballast one half of the deck with 60 tons of concrete to deflect it sufficiently to join it to the other half, rather than to jack up the span—to save time. A buckle appeared in the central plate that joined the two halves of the deck. The cause of the central plate buckle was being investigated when the 376-foot cantilever fell to the ground, without warning.

The international enquiry team into the collapse, organized by the British Government, was also given the responsibility for establishing the rules for

ABOVE: Milford Haven, Wales during the rebuilding work of the collapsed span.

ABOVE: West Gate Bridge seen here with one of the box girder spans completed.

LEFT: 1200 tons of steel buckle and fall 300 ft to the ground on October 15, 1970.

appraising steel box girder structures during construction, and for setting standards of safety checks and site procedures. In Britain traffic was restricted on all steel box girder bridges, pending a full structural investigation on each bridge, even though quite a few of them had been opened for several years. Mr Justice Barber of the New South Wales Supreme Court, who headed the enquiry team, concluded in his report:

> "While we have found it necessary to make some criticism of all the other parties, justice to them requires us to state unequivocally that the greater part of the blame be attributed to (the consultants) Freeman Fox and Partners."

The story of the West Gate collapse is about the complex relationship on site between the designer, the contractor, and specialist teams, and the lack of adequate standards and routines for checking and approving the prefabricated sections of the box girder during the construction.

Earthquake damage in the 1990s Volcanic eruptions apart, there is nothing on earth more frightening or devastating than an earthquake. In impact it is the equivalent of shock waves below ground of several atomic bomb blasts. Below the sea, an earthquake can give rise to tidal waves as high as 200 feet, which can submerge coral islands and wreak havoc with ships, harbors, and buildings some distance inland.

In recent years there have been significant earthquakes in Kobe, Japan, San Francisco, and Los Angeles that have collapsed bridges as well as buildings. The alarming concern for bridge engineers was that, having designed the bridge supports to resist the forces of an earthquake of a certain magnitude, they were shocked when many of their bridges collapsed. Assumptions about the most effective ways to absorb or resist the earthquake forces are still being studied and analyzed. No one can be certain that every eventuality will be catered for, but engineers in both Japan and California have come up with a new bridge-wrapping protection to the support columns and piers. They hope this will give the support added resistance against earthquake vibrations, but no one is absolutely sure.

It is a sad reality that it is only when something has been tested to the limit, and failed, that we can begin to fully understand the limits of the integrity of its construction, the behavior of the material it was built with, and the rigor of the structural analysis that the design was based on. With progress and change in our society continuing to evolve so rapidly, we may expect to stumble again on occasion.

ABOVE: The collapsed top carriageway of a double deck viaduct in San Francisco after the earthquake of 1989.

LEFT: "There is not much a bridge engineer can do to make a design completely safe against the devastating intensity of a big earthquake." San Francisco 1989.

6

Bridges and men

We often ask the same question about great ships, supersonic aircraft, classic cars, tall buildings, and bridges: Who built them? Sometimes we are given a clue to the identity of the designer by the name given to these artificial wonders—for example Bugatti, De Havilland, Rolls-Royce—but quite often they are named after the company that owns it, the city where it was built, or the river it crosses.

In the history of bridge building there have been a number of outstandingly brilliant designers, individuals who almost single-handedly changed the way bridges can be built to span farther, using daring and innovative concepts and new materials and assembly methods. Whether a bridge was built in a period when only stone and timber were available or a time when cast iron and wrought iron had been developed but not steel, or in an era when there were no computers or electronic measuring devices, just simple hand calculations, the rules of geometry, and the tape measure; certain individuals possessed a talent and skill far beyond those of their contemporaries. But their names do not appear on a bridge nor have any bridges been named after them. So who are they?

In this chapter the story of the great bridge engineers is told in the spirit of the prologue to David Steinman's book *Bridges and Their Builders* as a "heart-stirring

ABOVE: A precast segmental box girder section being lifted into position on one of the Florida Keys bridges.

OPPOSITE: The pedestrian walkway on the Brooklyn Bridge in New York is a popular place for tourists in summer.

narrative of high adventure and dramatic interest … an epic vision of courage, hope and disappointment … [for] a bridge is more than a thing of steel and stone: it is the embodiment of the effort of the human [mind], the heart and the hands."

Before beginning this tale of human endeavor let's just mention a few great men whose stories have not been told here because of space restrictions. Let's pay tribute to the past masters—the Roman engineer Caius Julius Lacer, builder of the Puenta Alcántara in Spain; to the leader of the French "Brothers of the Bridge" in the Middle Ages, Frère Benoit, a.k.a. St Bénézet, for building Pont d'Avignon; the English chaplain Peter Colechurch, who also belonged to a "Brothers of the Bridge" order, for building Old London Bridge; Taddeo Gaddi for the Ponte Vecchio; Antioni Da Ponte for the Rialto Bridge; Ammannati for the Santa Trinità; John Rennie for the New London Bridge; Isambard Kingdom Brunel for the Royal Albert Bridge at Saltash; Robert Stephenson for the Britannia Bridge; Gustav Eiffel for the Garabit Viaduct; James Eades for the St Louis Bridge; Robert Maillart for his concrete artistry and the Salgina Gorge Bridge; Gustav Lindenthal for the stunning steel arch of the Hell Gate Bridge; David Steinman for the Mackinac, the beautiful St John's and the Henry Hudson bridges; and Charles McCullogh for his "art deco" style of bridges.

And in recent times those on the short list toward immortality would include the bridge engineers Christian Menn, for his great teaching skill and aesthetic design of bridges; Jorg Schlaich for pioneering designs in cable-suspension bridges; the mercurial Santiago Calatrava for showing the world that modern bridge design can be an art form; Jiri Strasky for inventing the stressed-ribbon bridge; Michel Virlogeux for stretching the limits of the cable-stay bridge with the Pont de Normandie; and the bridge architects Alain Spielmann and Chris Wilkinson for putting architecture back into bridge design.

Now we will look in more detail at some of the great men who have populated the world of the bridge.

Jean-Rodolphe Perronet (1708–96)

Born in Suresnes outside Paris, Perronet died in Paris at the ripe old age of 88, in the happiest circumstances he could imagine, while supervising the completion of the Pont de la Concorde over the Seine, during the French Revolution. Perronet was the son of an army officer and when he was six his parents took him to the Tuileries, the French royal residence adjacent to the Louvre. He enjoyed a lucky childhood, for in Paris the young prince who was to become Louis XV was being entertained in an adjoining garden. The prince saw Perronet and asked him to come and play. Out of this chance meeting a lifelong friendship was born, which brought Perronet many special privileges and personal favors.

When young Perronet had grown into manhood he asked to follow in his father's footsteps, but rather than be an officer he wanted to train as a military engineer. No doubt he had been inspired by the great military engineers of the day, among them Jacques Gabriel, who had just founded the first government engineering department for the scientific advancement of bridge building—the Corps des Ponts et Chaussées. Plans of all roads, canals, and bridge works in central France had to be approved by this influential and prestigious body of engineers, who were all graduates of the Ecole de Paris.

However, Perronet's luck was to run out, since he was not admitted to the elite military engineering college; because he was not one of the three candidates selected of the many applications that were received that year. So he studied architecture instead at the Ecole de Paris and later started on a career building bridges. He studied trigonometry, algebra, architectural history, and mechanical science, using multiplication tables as his computer in order to work out the arithmetic of mass, area, and force.

Good drafting skills were essential in his profession and many hours were spent perfecting neatness and clarity of line drawings using the quill pen and leaded pencil, aided by only by a ruler and compass. By night he must have worked under the light of flickering oil lamps, sitting studiously hunched over a crude wooden table, wearing gloves and warm clothes in winter to keep out the cold.

During the first half of the eighteenth century in France, as Perronet was gaining experience as a bridge engineer, a number of important stone-arch bridges were built on the Seine and Loire rivers. Perronet assisted Hapeau, the chief engineer of the Corps des Ponts et Chaussées, whom Louis XV had appointed to design and supervise the construction of a new bridge at Blois when the famous old one was washed away in a flood. After Hapeau's death, Perronet supervised the completion of bridges that Hapeau had started over the Loire at Orleans and over the Seine at Mantes. He was an able administrator, a brilliant organizer and motivator of men, who was hardworking and earnest, with a work ethic that he expected others 6 follow.

It was while working on the Mantes Bridge that Perronet made the first of his momentous observations. When the first stone arch was nearly complete and the second was getting under way, he noticed that the pier between them was leaning slightly toward the unfinished arch. Many arches with a pier-width-to-span ratio of 1:5, like those at Mantes, stand quite safely when the bridge is completed, even if the piers probably swayed a bit during construction. This temporary condition was not considered a problem. But Perronet's curiosity as to why the piers should lean led him to conclude that the whole group of arches must provide support to one another when the bridge was complete, and that the thrust from each arch—which had caused the piers to lean during construction—was transferred to the abutments at each end.

What Perronet discovered was that the arches of these spans were not truly independent of each other. He was the first man to discover that the horizontal

ABOVE: Typical wooden framework for centering the stone arch. A drawing prepared by the office of Perronet.

thrust of the arches was carried through the arch spans and that the piers carry only the vertical load and the difference between the thrusts of the adjacent arch spans. By keeping the arch span the same there would be no thrust on the piers, so the piers could be greatly reduced in thickness. But the big problem was maintaining the stability of the pier during construction. Perronet's ingenious solution was to have all the arches in place before the centering was removed, and to build the arches simultaneously, working from the piers toward the crown to minimize any thrust on the piers.

With this discovery, there were suddenly two great advantages in building stone-arch bridges. The more slender piers would widen the navigable waterway and present less of an obstruction in the river for scour damage to the foundations. The arch span could be made flatter, by transferring the arch thrust back to the abutments, and in raising the haunches the arch could clear the waterway as much as possible. Perronet's ratio of pier width to arch span was about 1:10 and 1:12, compared with the customary 1:5 ratio in his day. And, instead of the elliptical three-segmental arch that Hapeau and Gabriel before him had designed, Perronet designed the arch as a segment of a large circle. It was an arch that was esthetically more pleasing than the changing radius of the segmental arch, prompting him to remark, "Some engineers, finding that the arches … do not rise enough near their springing, have given a larger number of degrees and a larger radius to this part of the curve … [Such] curves have a fault disagreeable to the eye."

The first bridge that Perronet designed was a single-span arch at Nogent on the upper Seine and, although he could not put all his theories to the test, the elliptical-arch span was a radical departure from other arches. A few years later he was asked to build a stone-arch bridge over the Seine at Neuilly, to the west of Paris. The time had come to show the world one of the most revolutionary creations in stone-arch bridge designs.

But as usual Perronet had to defend his ideas. Many prominent bridge engineers thought that the bridge would never stand up, that the King's money was being wasted, and that people would be killed building the bridge and using it if it ever did stand up. Perronet's nerve held out. He explained the principle of his design and how all the centering would be in place while the arches were being built and why the arches would not collapse once the centering was removed. His great reputation and persuasive arguments won through, but there were still many skeptics who thought the bridge would collapse and kept up their protest many years after it was finished.

The five arches of Pont Neuilly are each 120 feet in span, and the piers are 13 feet thick, making a pier-width-to-span ratio of 1:9.3, a slenderness that surpassed all other arches by a huge margin. To build the pier foundation across the river, huge cofferdams were built, which were drained by using a bucket wheel

operated by a paddle, driven by the river current. The river bed in the cofferdam was then excavated to a depth of 8 feet below low water level and then piled down to the gravel layer using a drop hammer which was driven by two horses. A raft of piles were driven to refusal and then cut off 10 feet below low water level. An open grillage of timber beams was laid across the piles with the interstices filled with stones cemented together with lime mortar. Following on, the masonry for the piers was built up from this platform.

Next, the timber centering for the arches was built from the pier foundations for each of the spans. Once the big masonry abutments at each end of the bridge

ABOVE: An artist impression of the Pont de la Concorde and the Place de la Concorde, commissioned for King Louis XV.

were finished, it was important to have all the arches finished in a single season between spring and fall, before the onset of the winter floods, which might wash away the centering. A total labor force of 872 men and 167 horses were engaged in transporting materials and building the arch voussoirs and spandrels for the bridge in the drive to get the arches completed. On September 22, 1772, in the presence of King Louis XV, all the centering was removed and the bridge was revealed. The Neuilly bridge was to stand in position for nearly 200 years, withstanding the floods and scour of the Seine, escaping damage from both World Wars. In 1956 it was demolished to make way for a larger bridge crossing. And so one of the most graceful and beautiful stone-arch bridges was lost to posterity in the name of progress!

Perronet was by now covered with honors and had became the first director of the Corps des Ponts et Chaussées, the first school of engineering in the world. The school and its first director, Perronet, secured the supremacy of French road- and bridge-building expertise throughout Europe. In the last 30 years of his life he was the chief of the Corps itself and publicly recognized as the premier engineer of the realm. But the culmination of his engineering achievement was not Pont Neuilly, nor his last bridge, the Pont de la Concorde: it was at Sainte-Maxence over the Oise, 35 miles north of Paris.

"Springing at a height of 18 feet from its tall and slender columnar piers (only 9 feet in diameter), it was truly a tour de force in stone arch construction," says Professor James Finch. It was the most slender and daring stone arch ever built, having a pier-width-to-span ratio of 1:12. The greater thrust from such a flat arch was carried by deep stone voussoirs, which occupied nearly all the space from the bottom of the arch to the roadway.

BELOW: "A great cheer went up when the centering was removed for Pont Neuilly Bridge in 1772."

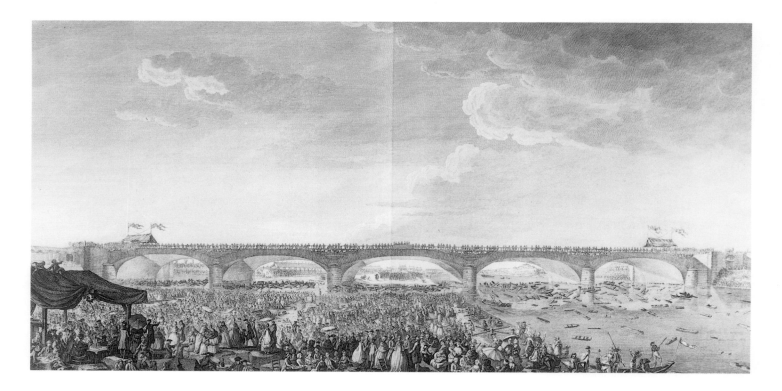

And just as daring in concept were the piers for the bridge. They consisted of a pair of double columns connected by a lateral arch. The pair of outer columns were connected by bracing walls, while the inner pair were joined by the lateral arch. The bridge was partially destroyed in 1814 by Napoleon's retreating army, and was destroyed when the Germans retreated from the Marne in 1914.

To celebrate the work of Jean-Rodolphe Perronet, the "Father of Modern Bridge Building," why not take a visit to Paris and see his last and most famous bridge, the Pont de la Concorde, built in the shadows of the guillotine during the reign of terror of the French Revolution. It was on this bridge that Perronet died in 1794 at the age of 88 while supervising the construction, living in a small hut built for him at the end of the bridge by the militia. Could he have been oblivious to the gruesome activity going on within earshot of the bridge, as the crowds cheered and drums rolled, when another victim was felled by the guillotine? He must have known many people close to the Royal household, some of whom would have been sent to the scaffold.

The beautiful irony of the Pont de la Concorde is also its charm: it was commissioned by Louis XV and finished by the people of the French Republic; it was called the Pont de la Concorde by Louis XVI and renamed Pont de la Révolution by the Republicans; and while chaos and terror was spreading through France, on the banks of the Seine a team of workers united under him, went quietly about their daily business sifting the rubble of the Bastille to salvage masonry to place on the great bridge.

DESSIN DU PONT PROJETTÉ SUR LA RIVIERE D'OISE, A PONT SAINTE MAXENCE

Thomas Telford (1757–1834)

An engineer is a person who was innocent of specialization and expected to be able to build or repair any structure, material or man-made work. Telford was such a person.

In 1777 Coalbrookdale, England, was the site of the first cast-iron bridge to be built. At the time this remarkable achievement did not attract great attention. Perhaps this was because it takes time for new ideas to become accepted, or perhaps it was because the bridge was an imitation of a masonry arch, with the cast-iron ribs taking the place of the stone voussoirs. But the second cast-iron bridge to be built some 15 years later and only three miles upstream from Coalbrookdale had quite a different reception. It was the first cast-iron bridge to express the true potential of the material.

Buildwas Bridge used only half the weight of material of Coalbrookdale and yet it was a flatter and more graceful arch. It was one of the many bridges that were to establish Thomas Telford as the greatest pioneer of the iron bridge structures in the world.

He was born in 1757 in the most humble of circumstances, into a poor family of hill farmers, living in a one-room cottage in the county of Dumfries in Scotland. He was brought up by his mother, because his father died when he was very

young. His mother relied on the goodwill of the neighbors to help with the small farm and to keep them from starvation.

More or less self-educated, with a cheerful disposition and a warm smile, the boy who kept sheep for relatives and ran errands for his neighbors was to become the greatest civil engineer in the country and the builder of the first iron-chain suspension bridge. After attending the village school he was apprenticed to a stonemason at age 14 and assigned to work on the estate of the Duke of Buccleugh in Langholm. There was plenty of work for young Telford, in building cottages, public buildings, dams, and bridges for the Duke, who was the richest landowner in all Scotland.

After serving his apprenticeship he left Langholm for Edinburgh, where the pay was better and there was plenty of work to do. Two years later at the age of 25, full of confidence in his ability as a stone-cutter and master mason, armed with letters of introduction from his Scottish patrons, he set off to find work in London.

Telford eventually secured work with the architect Sir William Chambers on the construction of London's famous Somerset House, starting as a lowly hewer but soon working his way up to become head mason. Telford's letters home to his mother give an insight into the mentality of his fellow workers and Telford's ambitions. According to Telford there were two distinct types of workers on the site—the majority of them just worked for their week's pay and a chance for a bit of fun after work, while Telford on the other hand was conscious of building a career and studying in his spare time. Telford from all accounts was a sociable, gregarious person with a lot of friends, but he never married.

In 1784 Telford left London to work on a project for the navy in Portsmouth Docks. Within a year he was promoted to general superintendent, and he then embarked on a long and challenging career as a civil engineer, first working as an engineer for the Ellesmere Canal Company, where he designed a series of aqueducts that included the first all-iron aqueduct at Longdon-on-Tern on the Shrewsbury Canal in addition to the famous Pontcysyllte Aqueduct. The Pontcysyllte remains unique in scale and magnificence, reminiscent of the grandeur of the Pont du Gard, and carries the Ellesmere Canal over the Dee valley near Llangollen in Wales.

ABOVE: Telford's engineering drawing for a later iron truss arch.

Nineteen cast-iron arches, each spanning 45 feet, 127 feet above the river bed, are carried on stone piers, to support the cast-iron aqueduct for 1,000 feet over the gorge. It was during this period that Telford settled in Shrewsbury, in the heart of the iron and coal industry, and watched with interest as Abraham Darby and John Wilkinson tried to work out the problems of building a cast-iron arch. He studied the plans for another tremendous cast-iron bridge that was being built by Tom Paine in America with a span of 400 feet over the Schuylkill river, not far from Timothy Palmer's covered bridge. Paine had the giant cast-iron ribs forged in Coalbrookdale but, shortly after they were formed, his financial backers pulled out and the project was scrapped. The iron was sold as scrap and was used for building an iron-arch bridge over the Wear at Sunderland in 1796.

RIGHT: Beautiful Craigellachie Bridge over the Spey in Scotland (1815).

Now a resident in Shropshire, Telford took up the position of County Surveyor of Shropshire on the personal recommendation of Sir William Pulteney, who had employed Telford while he was in Edinburgh and when he first went to London. As County Surveyor, Telford was required to design many roads and canals, and to work on bridge construction and even railroads. Where did all this new knowledge and expertise suddenly come from? Telford was always studying books and buildings whenever he could find the time—visiting old cathedrals, reading Vitruvius's books on architecture, studying Wren's designs and Perronet's books on bridges no doubt—on every related subject in the field of civil engineering. He had already helped to build masonry bridges while serving his apprenticeship in Scotland and had a good grounding in cast-iron construction, being so close to Coalbrookdale, and so it was natural for him to try his hand at a cast-iron bridge when the old bridge at Buildwas was washed away by a flood

ABOVE: Craigellachie Bridge, the first iron truss arch bridge.

in 1795. Buildwas Bridge was the first cast-iron bridge to show the world how the material could be used to greater advantage than masonry-arch construction. It was lightweight, spanned farther, and could be built out from each abutment without obstructing the waterway! Craigellachie Bridge, built in 1815 over the Spey, was Telford's masterpiece and the most beautiful iron bridge built by him; it was also the first metal-arch-truss structure that did not try to imitate the masonry arch and spandrel. Many famous bridges were built in England during the next two decades following Telford's mastery of cast-iron construction.

The career-minded Telford went north to design and supervise the building of the Caledonian Canal in Scotland to eliminate the laborious and perilous sea journey around the north cape of Scotland. Then in 1810 he was engaged in the Holyhead Road Survey, which aimed at improving communication between London and Ireland by developing a good road link between Anglesey (off North

Wales) and mainland Wales, for the new harbor at Holyhead. The great problem was bridging the Menai Straits, between Anglesey and the mainland.

John Rennie, one of Telford's great contemporaries, had suggested an iron bridge with a single span of 450 feet, but the cost of construction was prohibitive at £290,000 (approximately $470,000) in those days. Meanwhile Telford was being consulted on his ideas for a bridge crossing of the Mersey at Runcorn, and, having read about designs for suspension bridges in America and France, astonished his clients when he said to them, "I recommend a bridge of wrought iron, upon the suspension principles."

Telford knew that suspension bridges were notoriously flexible and unstable, although recent designs by James Finlay in America had been built with a level bridge deck and were meant to carry road traffic. But, even then, one of Finlay's bridges had collapsed under a drove of cattle and another under snow and ice. The suspension principle was simple in theory, it was cheap on materials, but it was more treacherous in practice than engineers in the 1820s realized. But Telford had the instinct to know that he could solve this flexibility problem.

Telford collaborated with a Captain Samuel Brown, who had become interested in suspension bridges, and had invented a new flat-iron chain which he believed capable of supporting heavy loads without deforming. They became good friends and worked together on a series of tests for a chain-link suspension bridge. The Mersey crossing did not go ahead because the cost was too expensive, but all was not lost. Much of the test work and research that Telford had carried out with Brown was to prove invaluable for Telford's biggest challenge.

The Menai Straits Bridge is Telford's great memorial, and secures his place in engineering history. And, though he was nearly 70 when it was completed, he continued on in his career almost to the day he died at the age of 77 in 1834. The

ABOVE: The 1000 ft long cast iron aqueduct of Pont Cysyllte is supported on brick arches.

BELOW: The Chirk Aqueduct built in 1801 carries the Ellesmere Canal over the Ceriorg Valley in Wales.

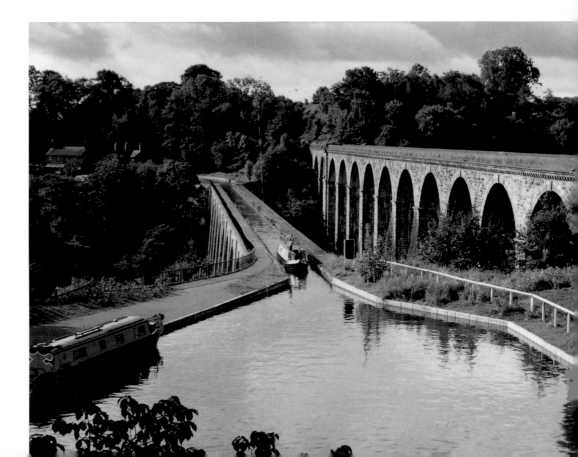

OPPOSITE: Pont Cysyllte aqueduct in Wales (1805), reminiscent of the grandeur of the Pont du Gard.

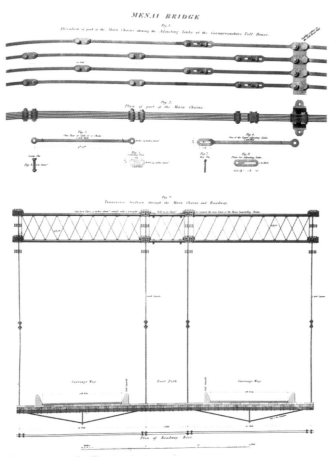

MENAI BRIDGE

bridge required 2,000 tons of wrought iron to build, which was a considerable amount, but it was far less than the 6,000 tons needed for the iron-arch spans of Vauxhall Bridge over the Thames in London. Through 1820 to 1822 work on two-pier foundations and towers continued, alongside testing of the flat-iron bars, 935 of which were needed to make each of the bridge's 16 cable chains. The two great suspension towers, which were hollow over the top third, were raised 30 feet above the roadway level and completed in the spring of 1824. The carriageway, which narrows to 9 feet to pass through the towers, was built in sections on the ground and fitted with cast-iron plates and saddles for connecting them to the chain cable. The rest of the year was spent getting preparations ready for suspending the first cable chain.

First a half-section of the chain on the mainland side was pulled over the tower and draped down the tower face, until it met a barge in the water. Then a portion of the chain was laid over the top of the Anglesey tower. Next a barge was floated out with the final section of the chain cable and one end was then attached to the existing chain while the other was fitted to heavy ropes pulled by a capstan on the other bank.

Telford's diary records the final moments: "... the said ropes passed, by means of blocks, over the top of the pyramid [tower] of the Anglesey pier. Then the workmen who manned the capstans moved at a steady trot, and in one hour and thirty-five minutes after they commenced hoisting, the chain was raised to its proper curvature, and fastened to the portion of chain previously placed at the top of the Anglesey pyramid."

Within 16 weeks the last of the 16 chains were lifted into place. The bridge deck, the toll gates, and side railings were all in place, and so, finally, was the road, with three thicknesses of fir planking, each layer separated by felt.

On January 30, 1826, the London mail coach occupied by Telford's assistant engineers, the mail-coach superintendent, and anyone else who could find a foothold on the coach crossed the world's first bridge over an ocean and the longest span in the world at 580 feet.

The astonishing feature of many of Telford's bridge structures was the relative ease of their construction, the lack of accidents, and the certainty that they could and would be built. Telford, by all accounts, was a good-humored man, an entertaining raconteur, an intuitive and brilliant engineer, and a lively writer, who kept a detailed diary of events and wrote regularly to his friends in Scotland and to his mother unfailingly until her death. He never married but in his letters there was often reference to a woman, an actress whose talent he admired. The explanation given by his close friend and biographer Rickman was simply "he lived as a soldier; always in active service!"

ABOVE: The Menai Straits Bridge as it appears today.

OPPOSITE ABOVE: A painting of the completed Menai Straits suspension bridge not long after it opened in 1826.

LEFT: "The first bridge to be built over an ocean" was also the longest span in the world at 580 ft when it was completed.

The Roeblings: John A., Washington A., and Emily Warren—a magnificent obsession

But a bridge is more than the embodiment of the scientific knowledge of the physical laws. It is equally a monument to the moral qualities of the human soul ... Let us then record the names of the engineers who have thus made humanity itself their debtor ... They are John A. Roebling, who conceived the project and formulated the plan of the bridge; his son Washington A. Roebling ... who directed the work from its inception to its completion upon the tragic death of his father ... and one other name which may not find a place in the official records, but cannot be passed over in silence ... this bridge is an everlasting monument to the self-sacrificing devotion of Mrs Emily Warren Roebling ...

So said the Honorable Abram S. Hewitt, civic leader of New York, during the inaugural speeches that marked the opening of the Brooklyn Bridge in 1874.

During the ceremony, a lonely paralyzed man, crippled and racked with pain, sat at an upper window of an apartment block in Columbia Heights, viewing the scene on the bridge through field glasses. Through them he saw the distinguished procession coming over the bridge, which included the President of the United States, the governor of the state, and many other notables, together with a glittering military escort.

His throat was choked with emotion, and he could barely hold back the tears, for this was the greatest moment of Washington Roebling's life. This day gave

RIGHT: Brooklyn Bridge. This picture was taken in 1916.

meaning to his suffering and his father's life and dedication before him. This was his father John Roebling's bridge, which he had built against all the odds.

Beside him at the window stood his wife, Emily, who had been his ministering angle through the years of pain and struggle that he had endured. She had been his eyes and his feet for ten harrowing years, recording his notes and carrying his every instruction to the workmen in the caisson, as he lay paralyzed from caisson disease (the bends). He could look back now with immense pride at the Brooklyn Bridge, at what he, his wife Emily, and his great father John Roebling had striven to achieve.

Suspension ideas John Roebling came to America in 1831 at the age of 25, not in search of God or bread, but to make a living as a civil engineer and bridge builder. Roebling was the son of a middle-class German family, with an excellent education at the Royal Polytechnic Institute of Berlin. He was traveling on a ship heading for Philadelphia with $400 in his pocket, which was the equivalent of a small fortune in those days.

Once landed, Roebling, his brother Carl, and his compatriots headed for the frontier lands of western Pennsylvania, to set up farmlands and to establish a community, which they called Saxonberg. Roebling kept up his engineering studies in the evenings while doing farm work in the daytime, and waited for an opportunity to get started as a civil engineer.

In 1837, through a good friend, he got a job with the Sandy and Beaver Canal Company. Two years later he was surveying the route of the railroad from

Harrisberg to Pittsburgh and that same year he had his first practical ideas on bridge building.

A feature of the canal system was the mechanism used to tow flatboats on rails, to the top of a steep hill or incline, and then put them back into the canal on the other side. Heavy hemp cables were used to haul the boats, but they frequently broke and on occasion caused fatal accidents. Roebling thought about using wire rope, an invention he had just read about in a German technical magazine. No one in America had ever *seen* a wire rope, let alone heard of its being used, and naturally the canal company did not think his idea would work.

Roebling decided there and then to set up his own wire-making company with machinery he could develop, but he needed people to make it. He called on his Saxonberg neighbors and willingly they turned themselves into an efficient team of workers.

Roebling's wire rope was stronger and easier to handle and outlasted the best hemp rope by an irresistible margin. With his civil-engineering skills he soon established himself as one of the leading canal engineers of the region, designing, building, and repairing aqueducts over rivers using suspension structures supported by wrought-iron cables spun by his company. The public and the technical press were much impressed with the daring suspension structures designed by the inventive Roebling, but he wanted to go on and build bigger structures, such as road and rail bridges.

He got his big break in 1851, when the mercurial Charles Ellet quit as engineer for the Niagara Falls Bridge, and Roebling was invited to take over.

RIGHT: Cincinnati-Covington Bridge, which was built in 1866.

Othmar H. Amman (1881–1967) Amman was born in Switzerland in 1881, in the village of Schaffhausen, where many years ago Hans Grubenmann built a famous truss bridge. Perhaps the timber bridge inspired the young Amman, who wanted to be an architect but was encouraged to study engineering at Swiss Federal Polytechnic Institute in Zürich, because he was gifted at mathematics.

He started work as a draftsman in an engineering firm in Germany, but soon became bored and frustrated, because his ambition was to be involved in building the biggest bridges in the world. The place to realize his ambitions was America, where the great Brooklyn, Quebec, and St Louis bridges had been built and where many more long-span bridges were being proposed.

He sailed for New York in 1904 and secured a job with a consulting engineer, Joseph Meyer. He then went to work with Frederick Kunz, whose Pennsylvania Steel Company was involved with the Queensboro Bridge in New York, a long-span cantilever truss. Then the next change came shortly after the Quebec Bridge collapse in 1907, when Amman offered his services, for no pay, to work with C.C. Schnieder, whose company was appointed to investigate the failure.

And it was in this roundabout way that the restless Amman met up with the famous Gustav Lindenthal, whose company were also advisers to the Canadian government on the Quebec Bridge collapse. At last he had found a practice that was to stretch his design talents, that suited his ambitions, and in Lindenthal he had a mentor he really admired. He worked for Lindenthal in all for ten years, with a break of a couple years when he was recalled to do army service in Switzerland during World War I.

When he first joined Lindenthal his first assignment was as first assistant in charge of office and field operations of the Hell Gate Bridge, which was about to begin construction. Unfortunately, Amman had to leave for army service in 1914, and did not see the Hell Gate finished. Upon Amman's return to America,

ABOVE: Othmar Amman aged 70.

BELOW: The Bayonne Bridge over Kill van Kull in New York, arguably the finest truss arch span ever built (1931).

Lindenthal placed him in charge of designing a railroad bridge across the Hudson at 57th Street in New York. In the end the project was not built because it did not attract enough sponsors, but Amman did not give up on the idea of building this much-needed bridge and thought that it might be lighter and cheaper to build if it was erected at some other point.

According to Margaret Durer Amman, Othar Amman's daughter, now aged 76, her father pointed out that a railroad bridge over the Hudson at its widest point was too expensive to finance. "He had suggested to Lindenthal that it would be better to build a bridge further upstream, roughly where the George Washington is today, and to make it suitable for motor cars and not a railway, as this was going to be the transport of the future. They argued over this for two years. No one was willing to put money into the scheme. In the end my father grew extremely frustrated and resigned. Governor Silver of New Jersey got to hear about my father's concerns and approached him about becoming Chief Engineer of the newly formed New York Port Authority," says Margaret Amman. "Of course Governor Silver and my father go back a long way, to a time when Lindenthal was nearly bankrupt, my father was penniless and I had just been born. You see after the Hellgate bridge Lindenthal had no more work, and slowly as the funds ran down he had to lay off every one of his staff including David Steinman, except my father, whom Lindenthal thought the world of." Lindenthal offered Amman a post of managing a clay mine in New Jersey, which he and Silver had bought at a knock down price, some years ago. It had been losing money. "The job came with a house and a big garden, so my father took it … there was no other work." Within two years Amman made that clay mine into a profitable pottery business and earned Lindenthal and Silver a small fortune for their minimal investment. Silver had not forgotten that.

Meanwhile, Amman continued to design other successful bridges for Lindenthal until 1925, when he left to become bridge engineer for the newly formed Port of New York Authority. The new public corporation was set up with the purpose of purchasing, constructing, and operating terminals and transportation facilities—which, broadly translated, meant lots of bridges.

In the course of the next six years, Amman, with the help of his assistant Alton Danna, was to design two magnificent, innovative, and record-breaking bridge

TOP: The truss of the Bayonne Bridge, New York.

ABOVE: Father and daughter (Margaret Amman-Durer) together in 1924.

spans, whose names are as famous as the cloud-busting skyscrapers on the Manhattan skyline. The record-breaking span of the mighty George Washington Bridge—a leap of 89 percent more than the previous record of the Quebec Bridge—was opened three weeks before the great steel-arch truss of the Bayonne Bridge over the Kill Van Kull in November 1931. The former was the longest span while the latter was the longest steel-arch span in the world. It was unprecedented that two structures of such scale had been undertaken and completed simultaneously by one organization and one man.

The Kill Van Kull waterway, which separates New Jersey from Staten Island in New York, is an estuary about a quarter of a mile wide. Through this waterway passes most of the cargo arriving and leaving New York, which in tonnage handled was greater than the Suez Canal at its peak. Any bridge that was proposed would need a clear span of 1,650 feet with no intermediate supports to obstruct or narrow the navigation channel.

ABOVE: Amman shaking the hand of Howard Hughes watched by Nelson Rockerfeller at the dedication ceremony when the second bridge deck was added to the George Washington Bridge.

LEFT: The tower of the George Washington Bridge (1931).

ABOVE: Goethals Bridge, New York (1929).

Amman's experience and knowledge of cantilever and steel-truss design was second to none and he also knew a lot about suspension bridges. From topographical survey maps of the areas, followed by borings and soundings, a thorough study of the ground conditions was prepared. This revealed that the estuary was underlain by good rock formation not too deep below low water level.

The choice for Amman was interesting, because he could design a cantilever-truss, suspension, or steel-arch bridge without difficulty for the clear span. In the end it was economic and aesthetic considerations that dictated the way forward. The first option, the cantilever-truss bridge, would be bulky and require more steel to construct than the other two, while the second option, the suspension bridge, would need deep and expensive excavations in the bedrock for forming the anchorages for the cables. The dense rock that made the suspension bridge expensive was in fact ideal to carry the large thrusts from an arch span.

Amman designed the Bayonne as a two-hinged, parabolic, through-arch truss, whose main ribs are the bottom chord of the arch and rise 266 feet from the water line. The top chord of the arch acts as a stiffener to the bottom chord. The central portion of the bridge deck below the arch is suspended from the trussed arch by wire rope hangers. The steelwork was supplied and erected by the American Bridge Company, who offered to use a new and cheaper carbon-manganese steel rather than nickel steel for the main chord. Amman ordered a series of tests on the new steel to check its suitability as a replacement material, and, although it was not as good in tension as the nickel steel, after some design modification it was accepted. Over 5,000 tons of it was used on the Bayonne Bridge, the first bridge to use carbon-manganese steel, which is the composition of modern steel.

The abutments that transmit the arch thrust directly to the bedrock are close to the water line, while the lightweight steel towers at the end of the arch span have been left unclad, just like the towers on the George Washington Bridge. Cass Gilbert, the architectural adviser to the Port Authority, detailed masonry to clad the towers, but it was omitted on economic grounds.

Controversy still surrounds the towers of the Bayonne. Would they have looked better clad in masonry, like the monumental towers of the Hell Gate and Sydney Harbour Bridges? Undoubtedly the bridge as it was built expresses the purity of its engineering design, and its lightweight construction. Cosmetic camouflage on this occasion would surely have been an aesthetic blunder.

Amman went on to design a number of other notable New York landmark bridges—the Triborough, the Throgs Neck, the Goethals, and the Outerbridge. But it was Amman's Bronx Whitestone suspension bridge, completed in 1939, that was universally acclaimed as the ultimate in suspension-bridge design. The slender plate-girder stiffeners for the deck span, and simple yet elegant towers, represented the very antithesis of nineteenth-century bridge construction. Just

BELOW: Verazzano Narrows Bridge, New York built in 1965.

OPPOSITE: Looking down the face of the tower of the Verazzano Narrows Bridge.

when Amman and his great rival David Steinman—who between them had designed most of the suspension bridges of North America—felt they had mastered the science of the long-span suspension bridge, Mother Nature dealt the bridge world a cruel lesson.

A year after the Bronx Whitestone Bridge opened, the Tacoma Narrows Bridge collapsed, causing bridge engineers the world over to re-evaluate their approach to suspension-bridge deck design. The Bronx Whitestone deck was rebuilt with a conventional heavy truss superimposed over the slim plate girders. Diagonal stays were also added, which did nothing to help the bridge aesthetics, while the Golden Gate underwent a $3,500,000 strengthening. Overnight, the heavy truss-girder bridge deck, which had been replaced by the slim plate girder, was brought back into suspension-bridge construction.

Amman's great rival Steinman beat him to win the commission for designing the big Mackinac suspension bridge in 1950, but Amman was to have the last laugh. In 1959 the Triborough Bridge and Tunnel Authority appointed Amman to design the colossal Verazzano Narrows Bridge, the longest span in the world. It was a scheme first proposed by Steinman in the 1920s, but after years of political debate he was never given the go-ahead.

It began two years after the Mackinac was completed and was the mightiest suspension bridge of all. Its size was matched by its huge expenditure: $325 million. No bridge had cost so much or used so much material before. The bridge was financed out of public funds and the cost recovered through car tolls, which has proved to be an extremely sound investment, since the traffic volume over the bridge doubled within eight years, going from 10 million to a staggering 25 million vehicles.

The tower rises 650 feet above the water line, the deck is suspended 226 feet above it, and the main span stretches a phenomenal 4,260 feet between towers, dominating the skyline and the waterway from Brooklyn to Staten Island. "Everything about the Verazzano Narrows Bridge," observed the *Engineering News Record* "is big, bigger, or biggest!" The four steel cables alone cost more than the whole of the Golden Gate Bridge, and there is enough strand in the cable to wrap themselves around the world a mind-boggling six times. The cables must support a dead weight of 120,000 tons from the bridge, plus the cable's own weight of 39,000 tons.

Despite the view that the Verazzano was not an outstanding bridge design, it is visually so impressive that even the heavy double-deck roadway looks in proportion when juxtaposed between the mass of the towers and the open span. It is a very photogenic subject, as demonstrated to great effect opposite, and is arguably one of the most impressive bridges that straddles New York's waterways. The year after the Verazzano Narrows Bridge was opened on September 22, 1965, Othmar Amman died at the age of 86.

Eugene Freyssinet (1879–1962)

> There is not any field of constructive activity—and I say not any, after close consideration—to which the idea of prestressing does not provide, often, unexpected possibilities.
>
> *Eugene Freyssinet*

I can best introduce Eugene Freyssinet to you by beginning this profile with an extract from an article called "His Own Man," written by Sir Alan Harris in *Concrete Quarterly* in 1992:

> What was he like? Small in stature, with a pussy cat face, he was capable of tigerlike roars, but his own angers were soon forgotten: all he asked for was complete devotion to the job! He had no taste for social life. At a celebration at the Ecole des Ponts et Chaussées to honor his 60 years, I found myself sitting next to him in the front row of a lecture hall faced by a stage full of the great and the good in French civil engineering, all waiting to speak in his praise. "Watch that lot," he whispered to me, "every time prestressing is mentioned, each looks as if he had a kick up the backside!" There was once a garden party at his house and it was a complete disaster, but he did turn up at my wedding in Paris and was charm himself.

The early years Freyssinet started as a junior in a local office of the Ecole des Ponts et Chaussées at Moulins, where his job was adviser to a number of rural mayors. He loved his work, knew his clients and their needs, and was given total freedom to design structures of his own devising, usually in reinforced concrete, which the mayors organized to have built by local labor. "Anyone," he said, "who told them that those bridges were contrary to the regulations would have run a heavy risk."

Many of those bridges are still standing today. The three identical bridges over the Allier just below Moulins are the most impressive. They were of three spans of 238 feet each, with very flat concrete arches. He built a test arch bridge using prestressed concrete as a foundation tie and discovered the existence of creep in

ABOVE: Eugene Freysinnet in his 50s.

RIGHT: Boution Bridge over the Alliers (c1912).

concrete. This discovery was to spark his early enthusiasm and later dedication to the art of prestressed concrete.

He first thought of it in 1904 but it took him another 20 years to bring prestressed concrete to full practicality. In ordinary reinforced concrete the liquid concrete is poured into forms or molds around the reinforcing bars. As the concrete hardens the formwork bears the load. Not until the forms are removed does the concrete come under any stress. This is the stage when the loading, combined with shrinkage as the concrete dries and temperature contraction as the concrete cools, does most damage and when cracks begin to appear in the tension zone, because concrete is weak in tension. Freyssinet's idea was to induce into the concrete, before erection or before the forms were removed, a precompression stress that would effectively counteract those damaging tension stresses in the finished structure.

Freyssinet received a special prize and much publicity when the second of the three bridges, the Pont Le Veurdre, was built in 1911. Suddenly he was a celebrity in the bridge world at the age of 30 and was receiving approaches from clients all over the world to design and build bridges for them. He resigned from state service and set up as a designer-contractor with two partners.

Then came World War I, when Freyssinet was seconded to the Army Engineer Works, where he explored and exploited the use of concrete, building shell-roof industrial buildings, many bridges, and a number of seagoing cargo ships made of concrete. By the end of World War I he was a master of concrete construction and a pioneer of concrete shell-roof construction and returned to his business to design and build all manner of concrete structures—from airship hangars and industrial complexes to record-breaking spans for concrete bridges that were cheaper than any other in the market—culminating in the innovative Plougastel Bridge over the Elorn. Plougastel's record-breaking concrete arches with three spans of 590 feet, which were the longest in the world at the time, and the unique method of fabrication and flotation of centering established Freyssinet as one of the greats among bridge engineers. It was while building Plougastel that Freyssinet found the answer to creep and how to design and build structures with prestressed concrete.

ABOVE: St Pierre du Vouvray, 1922.

ABOVE RIGHT: Centering for concrete arch of St Pierre du Vouvray.

BELOW: Prestressed beams of the Luzancy Bridge (1945).

A life of prestressed concrete Freyssinet was determined now to make prestressed concrete his vocation, and left his successful practice, much to the disappointment of his two partners, who thought he was mad to abandon everything for such a risky new venture. He was 50 and was prepared to devote all his private fortune to developing prestressed concrete.

After a period of waiting, a client emerged who wanted to supply 40-foot concrete pylons for power lines that would be needed for a major expansion of the electric power grid. The demand for power lines would be so large that Freyssinet and his backer would need to invest in five or six factories located around the country. Shortly after the pilot plant was set up and the first batch of the precast concrete hollow tubes with walls just under half an inch (or 10 mm) thick were produced, the market collapsed as the depression of 1933 swept through France and Europe.

His factory was sold for scrap and he was ruined. He and his wife were virtually penniless. And then he made the bravest gamble of his life. The Gare Maritime, the largest shipyard building in France, located at Le Havre, was suffering with disastrous settlement and sinking. The old quay walls were fine but not the building. The shipyard was the intended dry dock for building the SS *Normandie*, which was the finest and most beautiful ocean liner to have ever crossed the Atlantic.

To offer to remedy the problem was bold, and at the same time to fail would be financial disaster for an eternity. But what had Freyssinet to lose? His solution of jacking up the Gare Maritime building was accepted, though he had never used this technology before.

First he formed a prestressed concrete beam across the pile heads of the existing foundations. Then through holes made in the beams he used hydraulic jacks to push tubular-concrete piles, cast in lengths of about 6 feet 6 inches, down into the gravel layer, which commenced at some 33 to 50 feet below ground. When the concrete piles had reached a dense gravel layer, the subsidence slowed down.

He then used the piling jacks to restore the building to its intended level. Freyssinet was hailed as a great engineer and was congratulated by the leading civil engineers in France, including Edme Campenon of Campenon Bernard, who immediately engaged Freyssinet as their designer and championed his ideas for the rest of his life.

Prestressing was now established and, through the international reputation of Campenon Bernard, the stage was set for Freyssinet. Dams in North Africa were built, pressure pipes in concrete manufactured, many shell-roof structures and bridges were erected, precast bridge beams for mass production were granted licenses for manufacture in Germany and Switzerland.

Among the most important projects that were built was the family of five prestressed bridges—the Esbly, Annet, Trilbardou, Ussy, and Chagis—erected over the Marne between 1947 and 1950. According to Edme Campenon, this was the clearest souvenir of Freyssinet's creative genius.

In the first place, the choice of the structure—a raked leg portal bridge—connected with the obligation of leaving a wide navigable channel. Then the design had to fit between the profile of the carriageway and the soffit, which had a limited headroom. The construction method of placing the precast deck elements with an overhead runway, combined economy, speed of construction with a total absence of obstacles to navigation during erection. And finally the structure exerted thrusts on the abutments of the previous bridge which we were utilizing but which were far above those for which they were originally designed. Freyssinet found a way to re-use the old abutments by strengthening the soil behind the abutment by pre-loading the soil—it was soil prestressing.

RIGHT: Orly Airport Viaduct, one of the last bridges designed by Freysinnet.

Fritz Leonhardt (b. 1909) *In the 1950s while bridge engineers in America and Britain were preoccupied overcoming the problems of suspension bridges in high winds, in Germany a team of bridge engineers were developing a new type of suspension bridge—the cable-stay bridge. The leader of this group and the most honored bridge engineer living today, was Fritz Leonhardt.*

Leonhardt was born in Stuttgart in 1909 and enjoyed the discipline of school life at the excellent Dillman Gymnasium, from where he went on to study civil engineering at Stuttgart University. Leonhardt was an active and intrepid traveler in his youth, who would think nothing of walking from the northern rim of the Alps, down the length of Italy to Sorrento, in southern Italy, during his vacations. The beauty of the landscape, the magnificent mountain settings, and the unspoiled valleys were to have a lasting impression on him.

Much later, when he was a highly respected bridge engineer with his own practice, he would recall these memories in writing one of the great modern bridge books, entitled simply *Bridges*, in which he tackled the difficult subject of bridge aesthetics. It is a wonderful book, full of good ideas and sound advice and a bible on modern bridge aesthetics.

He won a scholarship as an exchange student in 1932 and went to Purdue University at Lafayette in central Indiana. It was a five-day sea journey on a White Star liner to New York. Having arrived in New York, Leonhardt lost no time in arranging to visit the two greatest living bridge engineers in America, David Steinman and the very elderly Othmar Amman. He already had the bug for bridges before he arrived in New York but this meeting with the great men, to discuss suspension-bridge design, was to inspire him for the rest of his life. As usual between semesters,

Leonhardt was off walking, and this time he managed to hitchhike his way all around the USA and down to Mexico.

On his return to Germany he got a job with the government Autobahnen (freeways) to design and build road bridges. His boss, Karl Schaechterle, was later transferred to the Berlin department, with the mission to improve the aesthetic design of bridges with the help of the famous architect Paul Bonatz. Leonhardt followed his boss to Berlin and had a "fruitful time." While there he met and married Liselotte.

RIGHT: The Cologne suspension bridge with its slim plate girder deck (1938).

BELOW: Knee Bridge, one of the first cable stay bridges to be built in Germany (1956).

"She was a wonderful person who gave birth to our six children," he reflects. "She had to overcome the grief of losing two of our lovely daughters early on. Liselotte was never jealous of my love for bridges and provided a harmonious home, where I could always gain strength in difficult times."

The freeway network was expanding at an alarming rate trying to keep pace with the industrial growth of cities along the Rhine valley. Germany had never needed to build a suspension bridge, since the steel truss and concrete arch were sufficient for most river spans. But the freeway demanded long elevated approaches above street level and the shoreline, and this required more ambitious bridging structures.

Early in 1938 Leonhardt was given the responsibility of designing Germany's first suspension bridge across the Rhine near Cologne with a clear span of 1,240 feet. He took great care to design an elegant bridge with good proportions, which was detailed with care. "Even the rivets were placed in good order," said Leonhardt. The slim, plate-stiffened bridge deck that he built was a departure from the heavy truss girder of the American designs. This innovative deck was later copied by Amman for the beautiful Bronx Whitestone suspension bridge and the ill-fated Tacoma Narrows Bridge. However, the plate-stiffened bridge deck had unfortunate consequences for suspension bridges when the deck was high above the water line, unlike the Cologne bridge where it was low. The high winds can cause the deck to oscillate at an alarming amplitude, and actually caused the Tacoma Narrows to collapse.

World War II interrupted Leonhardt's bridge work, when he was sent to Estonia to build factories to extract oil out of the shale deposits. At the end of it, with his

BELOW: Cologne-Deutz Bridge over the Rhine, an elegant steel plate girder bridge (1946).

home destroyed in the Berlin blitz, he and his family moved into a small house in the Black Forest, and he started up a new practice in Stuttgart.

Owing to the devastation of the bombing, there was a lot to do. The City of Cologne asked Leonhardt to rebuild the Cologne-Deutz Bridge over the Rhine—when completed it was the first slender, steel-box-girder bridge in the world. Then in 1952 he was asked to design ten Rhine bridges and a further five over the Mosele, which had been destroyed in the war. The Rhine bridges were among the first cable-stay bridges in the world. The Kniebrücke with a clear span of 1,050 feet was the second of Leonhardt's family of three cable-stay bridges at Düsseldorf, but they were not the first to be built.

The smaller Stromsund Bridge in Sweden, designed by Franz Dischinger, another German engineer, was first, finishing in 1955, a year before. It is difficult to know who was the first to conceive of the cable-stay bridge idea, because the cable-suspension principle was developed by human beings thousands of years ago. Probably the bridge that most anticipated the modern cable-stay bridge was the small concrete Temple Aqueduct, built over the Guadalete River in Spain by Torroja in 1928.

But there was no doubt that Leonhardt was regarded as the leading light, the modern guru of the cable-stay bridge. It is a more efficient structure than a suspension bridge—although it cannot span quite as far—because no large anchorages are needed at each end and it uses fewer cables to support the span.

The Kniebrücke designed by Leonhardt was to influence cable-stay construction for many decades to follow, with its cables arranged in an uncomplicated parallel or harp configuration. "I would like to acknowledge the contribution of René Walter and Jorg Schlaich, who developed the multi-cable stayed bridge to perfection," says Leonhardt modestly. "They developed the bridge deck to get more and more slender. Schlaich's Evripos bridge in Greece has only a deck thickness of 450 mm [18 inches] with no edge beam and a span of 215m [705 feet]—fantastic."

Equally fantastic was Leonhardt's Helgoland Bridge in Norway, which had a slim deck and span of 1,390 feet. It was worth noting that Jorg Schlaich learned his trade working for Leonhardt in his practice in Stuttgart for a number of years before eventually going on to establish his own consultancy.

After the failure of the Tacoma Bridge, Leonhardt worked on the development of an aerodynamic bridge deck to reduce wind oscillation. Wind-tunnel tests carried out on Leonhardt's prototype at the National Physics Laboratory (NPL) in England proved it had good wind stability. The aerodynamic deck behaves just like the reverse of an aerofoil, creating a net downward pressure in high winds and with no oscillations. When the Severn Bridge was being designed, the consultants Freeman Fox and Partners got to know about the wind-tunnel tests at the NPL and adopted Leonhardt's aerodynamic deck for the bridge.

The Severn Bridge was the first suspension bridge in the world to use a slender steel-box-girder deck, and has changed the way all suspension and cable-stay bridge decks have been designed since. Leonhardt's contribution to this innovation cannot be underestimated. Only Japanese bridge engineers still adopt the deep-truss-girder bridge deck based on the American model, although very recently even they have moved over to the aerodynamic deck for the first time, on the Kurushima suspension bridge.

Fritz Leonhardt has worked in many other fields of engineering besides bridge design: his practice, Leonhardt Andra and Partners, designed the Munich Olympic

ABOVE: Aesthetics was an integral part of Leonhardt's design philosophy, exemplified here in the Kochertal Viaduct near Gieslingen, Germany (1979).

BELOW: The aerofoil deck section of the Kurushima suspension bridge.

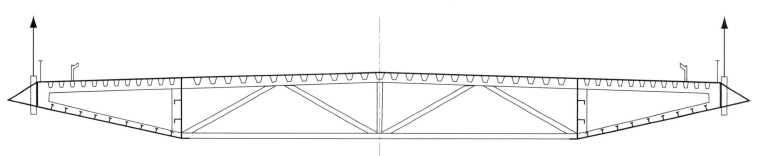

Stadium cable-net roof canopy with the architect Frei Otto; he was the partner responsible for over 200 telecommunications towers, designed in both concrete and steel, which include the Stuttgart TV Tower, Hamburg, Mannheim, Cologne, and many others. He has designed bridges in Brazil, Venezuela, India, the USA, and Japan and has acted as technical adviser to governments on numerous major bridge schemes. He has been a professor of engineering at Stuttgart University for many years and was the rector there for two years between 1967 and 1969. He has received many international honors and decorations for his achievements, including an honorary doctorate from the University of Bath in England and the Gold Medal of the Institution of Structural Engineers in London.

Leonhardt, now approaching his 90s, will be best remembered for an epistle on bridge aesthetics he wrote in his book, *Bridges* — a discipline he feels has been neglected in modern bridge design.

The esthetic qualities of the built environment have a profound influence on the human condition and the health of a society. Bridges are an essential part of the built environment. Massively, crudely shaped bridges, particularly freeway overhead bridges can destroy an environment, and create ghettos where crime and poor health flourish. Bridge esthetics should and must be the aim of every bridge that is designed or built in the world today. Artistic advice from suitably qualified architects on the design of bridges should not only be cultivated in the future, but also actively encouraged.

Jean Muller (b. 1925) "A resolute desire to simplify form"

That phrase was to stay in the subconscious of Jean Muller for the rest of his life, and to influence his approach to bridge design and civil engineering. It was spoken to him in 1952 by his boss, Eugene Freyssinet, a genius whose creative engineering skill has very obviously passed from master to pupil, when you consider the vast array of Jean Muller's accomplishments. There are the well-known and not so well-known pioneering bridge structures, barrel-vaulted dams in Africa and the Middle East, nuclear reactor plants in France, concrete-hulled supertankers, oil platforms, and harbor works—many of these structures were the first in their field; several were patented.

"His mind is like a volcano, and he has the energy of a twenty-five-year-old," says Jean Marc Tanis, chief executive of Jean Muller International (JMI). "He is unique and is still innovating, coming up with brilliant ideas on how to build or design a structure economically. Recently we asked his advice on a bridge we were designing for an international bridge competition in Malaysia and he came up with the most fantastic solution for a long-span cable-stay structure. We won the competition."

Born in Suresnes, a suburb of Paris, in 1925, Muller was one of five children and the middle one of the group. He followed his older brother Jacques into the construction industry and worked with him when he joined Campenon Bernard. His brother stayed on for 40 years, unlike Jean, who went on to form his own practice. His father worked in the largest bank in France, the CCF, in the foreign department, while his mother looked after the family and organized family gatherings and party games.

"My mother had the greatest influence on me, teaching me moral values, and social etiquette," Muller recalls fondly. "She was very strict with us and we could not go out late at night nor meet with girls. She encouraged my fondness for music and the piano. Playing the piano was a very rewarding pastime when I was growing up during the war years. I could play quite well."

At school he preferred physics and mathematics, and did not like the written subjects like literature and history. In 1944 he entered the Ecole Centrale des Arts des Manufactures for a four-year degree course in mechanical science, specializing in his final year in construction rather than the mining or chemical-engineering options. Immediately after graduating he went to work for STUP, the firm of Eugene Freyssinet. "The president of STUP was

ABOVE: Boulounais Bridge, France (1997).

LEFT: Jean Muller with a portrait of his mentor Eugene Freyssinet.

ABOVE: The flat steel arch of Nemours Bridge over the Grande Canal du Havre, Normandy, France (1995).

RIGHT: A typical precast concrete segmental section.

introduced to me while I was at university and encouraged me to take up construction in my final year. I had also heard a lot about Freyssinet and was very attracted to construction," says Muller.

He spent nearly five years with STUP, and worked on a number of civil-engineering projects, enjoying his time in Venezuela designing the Caracas Bridge, a large concrete-arch structure, and later on in New York, where he was sent to start up a new office.

It was while he was in New York promoting prestressed concrete that Muller developed the idea of match casting—precast box-girder sections.

"The Shelton Road Bridge just north of New York was not a huge span, but it was too long to precast and to transport to site in one piece. So I decided to cast the box-girder beam in three pieces. To make sure we had a precise fit on site, we match-cast the adjoining sections using the face of the first piece as the mold for casting the next pieces, rather like making a jelly from a jelly mold," Muller explains.

The sections were taken to site and assembled on a scaffold, where the butt joints were glued with epoxy resin and then stitched with prestressing cables threaded through the units. The beam was then stressed and, after grouting up the cable ducts, the scaffolding was removed. To this day it looks pretty good, with no signs of corrosion or spalling to be seen. Not bad for a bridge that was built in 1952.

This was the prototype for later match-cast, glued, segmental bridges designed by Muller and copied by bridge engineers the world over. Freyssinet had suggested this idea to him during a private conversation years before the Shelton Bridge, and as Muller recalls, "You don't forget a private discussion with Freyssinet: he does the thinking and talking while you listen."

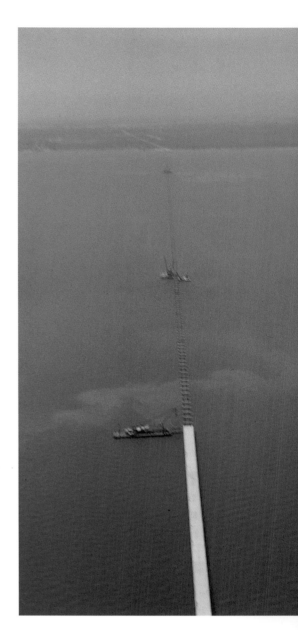

ABOVE: Ponchartrain Bridge under construction, New Orleans, USA.

LEFT: The erecting procedure for the Linn Cove viaduct.

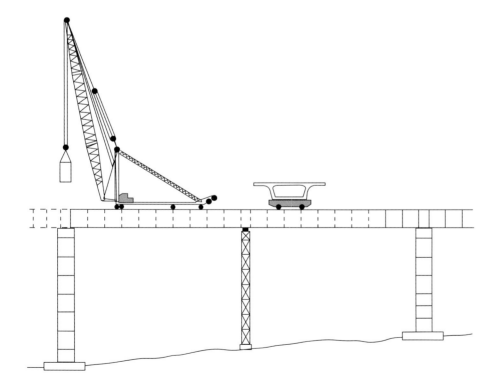

ABOVE: The Confederation Bridge across the Northumberland Strait, Canada (1997).

TOP RIGHT: Linn Cove Viaduct, North Carolina, USA (1982).

RIGHT: Launching the box girder span of the Linn Cove viaduct.

Before he left the USA to return to France in 1955, Muller worked on the construction engineering for building the Louisiana Rail Bridge over Lake Pontchartrain, west of New Orleans. It was 23 miles long, had 2,170 spans of precast, prestressed concrete, and is the second longest bridge in the world.

He was made chief engineer of Campenon Bernard, then technical director, and finally promoted to scientific director of the group in 1976. During this time he was responsible for the development of bridge-building techniques using concrete segmental construction, and worked on oil platforms and even proposed a supertanker with a concrete hull during the Suez crisis. He is proud of the 426-foot-long, multispan Choisy-le-Roi Bridge over the Seine just upstream of Paris, built in 1962, with its glued-segmental, balanced-cantilever construction and long-line method of precast production—the first time this method was used anywhere in the world.

Muller also spent quite a lot of his time with Campenon Bernard designing large multi-arch dams. In Algeria at one stage when he was designing the Erraguene Dam he was caught up in the a civil war when the Feligas rebels tried to overthrow the government. "I kept my head down and tried to keep my distance from the bullets and mortar bombs that were being fired in the capital."

TOP: Lowering a box girder segment into position for the Linn Cove viaduct.

ABOVE: Choisy-le-Roi Bridge over the Seine (1962).

OPPOSITE ABOVE: Brotonne Bridge near
Rouen, France (1977).

OPPOSITE BELOW: The Sunshine Skyway,
Tampa, Florida, USA (1986).

Of those that he designed he remembers vividly the Latyian Dam in Iran close to Tehran, with the beautiful snow-capped mountains in the background; Sourlages Dam in France with its barrel-vaulted arch and two great buttresses; the Sefidad Dam in Iran, a huge gravity dam 360 feet high, which needed over 1 million cubic meters (28,250,000 cubic feet) of concrete; but the best of all was the Djin Djin Dam in Algeria, built among the ruins of the Roman Empire. Then there was a period of designing prestressed-concrete containment structures for nuclear reactors when Europe went nuclear for new energy sources. The Tiage reactor in France was the first to use heavy water as the moderator, with natural uranium as the catalyst. It was a cleaner and safer system than Westinghouse's enriched-uranium rod process, but it never caught on outside France. Muller's team designed the prestressed concrete pressure vessel for the nuclear reactor.

When the demand for nuclear power stations dwindled, Muller went back to the USA to see if he could secure bridge contracts to keep the technical team at Campenon Bernard fully employed. A chance meeting with Eugene Figg in 1978 led to an agreement to combine resources to tender for the really big bridge projects that were coming up in the South. And thus Figg and Muller Engineers was set up, jointly owned by Campenon Bernard, Muller, and Figg.

BELOW: Sallingsund Bridge, Denmark (1974).

They took the US bridge market by storm. The long, multispan Florida Keys bridge was followed by the ingenious Linn Cove Viaduct cutting a swathe through the forested slopes of the North Carolina mountains. Muller devised a span-by-span, gantry-launched assembly of the precast box-girder beams, and the building of the next pier support from the finished span in order to cause the minimum disturbance to the fauna and flora of this beautiful region. "Where possible we tried to save as many trees as possible: only those on the direct line of the viaduct had to be cut, but no more, as we eliminated the need for site-access roads during span erection by using the completed span itself," he recalls.

When Linn Cove was finished it seemed as though the bridge had been dropped into place from the sky. There were very few scars and telltale spots to show where any excavation, temporary works, and construction traffic had been.

Then, in 1986, came the landmark bridge of the state of Florida and one of the greatest modern bridges, the cable-stayed "Sunshine Skyway." It's an attraction in its own right. "I designed the concept myself, the pronounced slope of the center span, the shape of the towers, the cable fans arrangement and so on … There were no architects," says Muller, quite modestly.

After this great success Figg and Muller disbanded, and Jean Muller went on to establish JMI, Jean Muller International, a world-class bridge design consultancy with offices in San Diego, Tallahassee, Paris, and Caracas. Under the JMI flagship emerged the highly acclaimed, cable-stayed Isère Bridge in Romans, regarded as one the Europe's finest bridges; the externally prestressed-concrete

Long Keys Bridge, a decisive breakthrough in new technology, which was the precursor to many similar bridge spans throughout the world; and further refinement of gantry-launched and segmental construction with the H3 viaduct in Hawaii and then the Prince Edward Island Bridge across the ice floes of Canada.

Eugene Freyssinet was already using prestressed-concrete segments for constructing the famous Marne bridges in the late 1940s when Jean Muller joined his practice. Since then, Muller has continued to promote this process to its current state in the art. He evolved match-cast segmental joints, the first launching gantries, and external prestressing for box-girder segments, and was the first to use precast segmental construction for a cable-stay structure on the Brotonne Bridge. But that is not all: very recently at the age of 64 he schemed the preliminary design for a massive long-span, cable-stay bridge in Malaysia—it is yet another breakthrough in cable-stay construction, allowing spans of around 3,300 feet to be built.

"It is more efficient than the record-breaking Pont de Normandie," says Muller, "because it needs only one plane of cable strands, rather than two, and incorporates a stiff but lightweight composite deck over the central span, where the top slab section is prestressed concrete, but the section below is a tubular-steel truss." It will be the longest cable-stay bridge in the world when it is built.

A bridge in the hands of a genius like Jean Muller is not just a utilitarian structure: it becomes a work of engineering art.

OPPOSITE ABOVE: Chillon Bridge, Switzerland (1966).

OPPOSITE BELOW: Pont Isère, Romans, France (1991).

LEFT: CAD image of the record breaking Sungai Johore, also known as "The Ceremonial Bridge" in Malaysia, now under construction.

BELOW: Roize Bridge, Grenoble, France (1992).

Glossary

Abutment: the end support of an arch bridge, built on the bank, which resists the horizontal thrust of the arch(es).

Aerodynamic deck: a bridge deck with an aerofoil cross-section which provides aerodynamic stability in high winds.

Air spinning: a method of making the cables for a suspension bridge. The individual wires are passed back and forth across the entire span, until the suspension cable is built up to the required diameter.

Anchorage: the support structure for anchoring and holding the cable ends of a suspension bridge.

Arch: a curved bridge span where the weight and forces acting on the span are carried in compression outward to the supports.

Aqueduct: a bridge or viaduct carrying a canal or channel of water across the span(s).

Barrel Vault: an arch of semicircular shape, often of brickwork or masonry, whose length is longer than its span. Sometimes called a cylindrical vault.

Bascule: a type of drawbridge with a counterbalance weight at the ends, which causes the span to rise and open when it is engaged.

Beam: a narrow and deep, or shallow and wide, rectangular member or an I-section or T-section structural element, spanning between pier supports. A bridge is often made up of number of these beams. Also known as girders when they are deep.

Bedrock: the solid rock layer beneath the silt and sand of a river bed or estuary.

Bends: see **Caisson disease**.

Bowstring arch: an arch whose ends are linked or tied together to resist the outward thrust.

Box girder: a deep hollow box beam which can have a rectangular or trapezoidal cross-section.

Cable-stayed bridge: a bridge whose deck is supported directly by a series of inclined cables connected to the pylon or mast and not from the vertical hangers of a suspension bridge.

Caisson: a structure for keeping water out while deep foundations are being excavated. In the lower working chamber water is kept out by increasing the air pressure.

Caisson disease: a disease that can affect workers working under compressed air who come too quickly out of the airlock. It is caused by bubbles of nitrogen coming out of the blood. Also called decompression sickness or the bends.

Cantilever: a structure or beam that is unsupported and free at one end, and fixed and supported at the other.

Cast in place: concrete that is formed in its final position by placing wet concrete into formwork or shuttering and allowing it to harden: it is the same as *in-situ* concrete.

Cast iron: a brittle alloy of iron with a high carbon content that is good in compression and weak in tension.

Catenary: the sag or profile of a rope or cable suspended from two points, as on a suspension bridge.

Cement: a powder that, when mixed with water, binds a stone-and-sand mixture into a strong concrete within a few days.

Centering: a temporary framework of timber to support the masonry while an arch is being built.

Chord: the top or bottom horizontal part of a truss.

Cofferdam: a structure for keeping out water to allow excavations and building of foundations in ground below water level. It differs from a **caisson** (q.v.) because it is open to the air.

Compression: a force that tends to shorten and compress something.

Concrete: a mixture of sand, stone and water bound by **cement** (q.v.), which hardens into a rocklike material.

Corbeling: successive layers of masonry or brick projecting beyond each other.

Creep: the slow permanent deformation of a material under stress, a characteristic of concrete.

Crown: the highest point of an arch.

Cutwater: the end of a pier base that is pointed to cleave the water.

Deck: generally the top side of a beam, box girder, or truss which forms the running surface for vehicles or pedestrians.

Decompression sickness: see **Caisson disease**.

Elliptical arch: an arch with a curve that becomes much tighter toward the crown.

Empirical formula: a formula or design rule based on one or many series of observations but with no theoretical backing.

Eyebar: the unit from which the chains of a suspension bridge were constructed, with a flattened ring at each end for linkages.

Falsework: temporary support scaffolding used during construction.

Fan configuration: the arrangement of cables on a cable-stay bridge which fan out from the pylon just like the folds of a paper fan do when it is open.

Flange: the top and bottom plates of a box girder, plate girder, or an **I-beam** (q.v.).

Formwork: temporary timber or metal shuttering to contain and support concrete while it hardens.

Girder: a large or deep **beam** (q.v.).

Hangers: the wires or bars from which the deck is hung from the cables in a **suspension bridge** (q.v.)—also known as suspenders.

Harp configuration: a cable-stay arrangement where the stays radiate out parallel to one another like the strings of a harp.

Haunch: part of the arch between the springing and the crown.

I-beam: a beam or girder with an I-shaped cross-section.

In-situ concrete: see **Cast in place**.

Keystone: the voussoir at the crown of the arch.

Navigation span: the part of a bridge with maximum clearance for shipping.

Ogival arch: a pointed arch.

Pier: the support of a bridge deck span, that is not on the bank. It is also a general term used for the base or foundation of a bridge.

Plate girder: a flat bridge deck with a shallow rectangular section.

Pointed arch: an arch with an angle at its crown.

Pontoon bridge: a bridge formed from boats, logs, or drums floating on the water and tied together.

Portal: a frame with uprights connected by a horizontal member at the top, just like the goal posts in soccer.

Precast concrete: concrete that has been hardened, taken out of its formwork or shuttering, then transported to the site and placed in position.

Prestressed concrete: steel wires or strands within the wet concrete are stretched (pretensioned) and held fast until the concrete has hardened. The tension in the wires keeps the concrete in compression and gives it a greater tensile strength. Alternatively, the strands are threaded through plastic or metal tubes in the hardened concrete and after the strands are tensioned (post-tensioning) they are anchored and grouted into place.

Pylon: the vertical mast or tower above the bridge deck to which the cable stays are fixed.

Reinforced concrete: concrete that is strengthened with high-tensile steel bars or rods to give the concrete tensile strength and ductility.

Segmental arch: arch formed from the segment of a circle.
Semicircular arch: arch forming a half-circle.

Shrinkage: the shortening of concrete that occurs as it dries and hardens.

Side span: the outer or end spans of a suspension bridge, from the tower to the anchorage, balancing the central suspended span.

Side sway: the side-to-side movement of a bridge deck in a wind.

Spandrel: the area of an arch above the **voussoirs** (q.v.) and below the bridge deck.

Springing: the point where the end of an arch meets an abutment or a pier.

Steel: an alloy of iron which has more carbon than **wrought iron** (q.v.) and less than **cast iron** (q.v.), combining the tensile strength of one and the compressive strength of the other.

Stiffening truss: a truss usually beneath the entire deck of a **suspension bridge** (q.v.).

Striking: the action of removing formwork in concrete and centering from beneath a completed arch.

Suspension bridge: a bridge with its deck supported by large cables or chains draped from towers.

Tensile strength: the ability of a material to withstand tension.

Tension: a force that tries to stretch and lengthen something.

Torsion: the stress produced when a structure is being twisted

Tower: the vertical support structure of a suspension bridge from which the cables are hung.

Truss: a frame of tension and compression members which together make up a long-span beam.

Viaduct: a bridge carrying a road or a railroad.

Voussoirs: the wedge-shaped stones from which an arch is formed.

Web: the side plates of a box girder or the vertical plate of an I-beam.

Wrought iron: iron that has been hammered into shape, with a low carbon content, low compressive strength, and high tensile strength.

Zigzag suspension: the arrangement of suspension hangers in zigzag fashion rather than vertical for added wind stability, and first introduced on the Severn Bridge, which links England with Wales.

Bridge facts

Longest bridge in the world

The longest bridge in the world is a viaduct 90 miles long in the Hwang Ho Valley in China, followed by the Lake Pontchartrain trestle bridge near New Orleans, which is 23 miles long. In Europe the longest bridge over water for many years was the Oosterschelde in Holland with a length of 3.4 miles. It has recently been beaten by the Oresund crossing linking Denmark to Sweden, with a bridge length totaling 8.3 miles, and which includes the record-breaking suspension span of the East Bridge.

Largest bridge-building program

The largest civil-engineering undertaking and the most expensive and colossal bridge-building program ever planned is taking place in Japan as I write this. There are no fewer than 17 major bridges being built—each one in itself would be a major undertaking for any other country in the world, but not Japan. This multibillion-dollar program will run for 15 years, and will see some of the longest and tallest cable-stay and suspension bridges ever built. The bridges will link the mainland of Honshu to the island of Shikoku via three major arterial highways—two of which carry railroads—across the inland sea of Seto. To the south will be the Onomichi Imabari highway,

with a total of nine bridges, one of which includes the Tartara bridge, the longest cable-stay span in the world, and the three elegant Kurushima suspension bridges. To the north is the comparatively short Kobe Naruto highway, which has

only two big bridges, but one of them is the Akashi Kaikyo, the longest span in the world. In between these two highways is the Kojima Sakaide route with a mixture of cable-stay and suspension bridges, totaling six in all.

The record for the longest bridge span

Name	Year	Location	Type	Span in feet
Trajan's Bridge	104 AD	Danube River	timber arch on stone piers	170
Trezzo	1371	Italy	stone arch	236
Wettingen	1758	Germany	timber arch	390
Menai Straits	1826	Wales	chain suspension	580
Fribourg	1834	Switzerland	wire-cable suspension	870
Wheeling	1849	West Virginia	suspension	1,010
Lewiston	1851	Niagara Falls	suspension	1,043
Cincinnati	1867	Ohio River	suspension	1,057
Clifton	1869	Niagara Falls	suspension	1,269
Brooklyn	1883	New York	suspension	1,595
Firth of Forth	1889	Scotland	cantilever	1,700
Quebec	1917	Canada	cantilever	1,800
Ambassador	1929	Detroit	suspension	1,850
George Washington	1931	New York	suspension	3,500
Golden Gate	1937	San Francisco	suspension	4,200
Verazzano	1965	New York	suspension	4,260
Humber	1981	England	suspension	4,624
East Bridge	1998	Denmark	suspension	5,328
Akashi Kaikyo	1998	Japan	suspension	6,529

Bridge type and longest spans

Type	Span in feet	Year	Location
Cantilever truss	1,800	1917	Quebec, Canada
	1,710	1890	Firth of Forth, Scotland
	1,644	1974	Commodore John Barry, Delaware, USA
	1,500	1943	Howrah, Calcutta, India
Steel arch	1,700	1981	New River Gorge, West Virginia, USA
	1,675	1931	Bayonne, New Jersey, USA
	1,670	1932	Sydney Harbour, Australia
Concrete arch	1,280	1979	KRK, Zagreb, Yugoslavia
	1,000	1964	Gladesville, Sydney, Australia
	951	1964	Foz-do-Iguaco, Brazil
Steel box girder	984	1974	Rio de Janeiro, Brazil
	837	1956	Sava, Yugoslavia
	831	1966	Zoo, Cologne, Germany
Cable stay	2,919	1998	Tartara, Hiroshima, Japan
	2,808	1995	Normandie, France
	1,984	1996	Quingzhou Minjang, China
	1,974	1993	Yangpu, China
Suspension	6,529	1998	Akashi Kaikyo, Japan
	5,328	1998	East Bridge, Denmark
	4,625	1981	Humber Bridge, Hull, England
	4,260	1964	Verazzano, New York, USA
	4,200	1937	Golden Gate, San Francisco, USA

Useful references

Brown, David: *Bridges—Five Thousand Years of Defying Nature*, Mitchell Beazely, UK, 1993

Delony, Eric: *Landmark American Bridges*, ASCE, USA, 1993

Gies, Joseph: *Bridges & Men*, Cassell & Co, UK, 1963

Hayden, Martin: *The Book of Bridges*, Marshall Cavendish, UK, 1976

Kingston, Jeremy: *How it is made: Bridges*, Threshold Books, UK, 1985

Leonhardt, Fritz: B*rücken (Bridges)*, the Architectural Press, UK, 1982

Plowden, David: *Bridges—The Spans of North America*, Viking Press, USA, 1974

Salvadori, Mario: *Building—from caves to skyscrapers*, Atheneum, USA, 1979

Steinman and Watson: *Bridges and Their Builders*, G.P. Putnams, USA, 1941

Wittfoht, Hans: *Building Bridges*, Beton Verlag, Germany, 1984

Useful Sources of Information

Caltrans, District 4 Public Affairs, 111 Grand Avenue, Oakland, CA 94612, USA

Ecole Nationale Des Ponts et Chaussées, Centre de Documentation, 6–8 avenue Blaise Pascal, Cité Descartes, Champs sur Marne, 77455 Marne La Vallée, Cedex 02, France

HAER, US Department of the Interior, National Park Service, 1849 C Street, NW, Washington, DC, 20240, USA

ICE Library, The Institution of Civil Engineers, 1 Great George Street, Westminster, London SW1P 3AA, UK

MTA, Public Affairs, Bridges & Tunnels, 18th Floor, 10 Columbus Circle, New York, NY 10019–1203, USA

New York Port Authority, Customer Relations, Tunnels, Bridges & Terminals, One World Trade Center, New York, NY 10048, USA

Index